Videoconferencing and Videotelephony

Technology and Standards

For a complete listing of the *Artech House Telecommunications Library*,
turn to the back of this book.

Videoconferencing and Videotelephony

Technology and Standards

Richard Schaphorst

Artech House
Boston • London

Library of Congress Cataloging-in-Publication Data
Schaphorst, Richard.
Videoconferencing and videotelephony: technology and standards/Richard Schaphorst.
p. cm.
Includes bibliographical references and index.
ISBN 0-89006-844-5
1. Interactive multimedia. 2. Videoconferencing. 3. Digital communications. I. Title.
QA76.76.I59S32 1996
621.39'9–dc20 96-19494
CIP

British Library Cataloguing in Publication Data
Schaphorst, Richard
Videoconferencing and videotelephony: technology and standards
1. Videoconferencing 2. Video telephone 3. Videoconferencing - Standards 4. Video telephone - Standards
I. Title
621.3'85

ISBN 0-89006-844-5

Cover design by Kara Munroe-Brown

International Standard Book Number: 0-89006-844-5
Library of Congress Catalog Card Number: 96-19494

10 9 8 7 6 5 4 3 2 1

Contents

Acknowledgments xi

Chapter 1 Introduction 1

Chapter 2 Video Teleconferencing: Benefits and System Design 5
2.1 Introduction 5
2.2 Benefits of Video Teleconferencing 5
2.3 Video Teleconferencing Overview 7
2.4 Video Teleconferencing Room Considerations 10
2.4.1 Room Size, Furniture, and Location 10
2.4.2 Cameras 12
2.4.3 Display System 13
2.4.4 Audio 14
2.4.5 The VTC Controller 16
2.4.6 Administration 18
2.5 Quality and Performance Considerations 18
2.5.1 General 18
2.5.2 System Considerations 19
2.5.3 Picture Format 19
2.5.4 Common Intermediate Format 20
2.5.5 QCIF 20
2.5.6 Frame Rate 21
2.5.7 Encoding Algorithm 21
2.5.8 Bit Rates 21
2.5.9 Audio 22
2.5.10 Bandwidth 23
2.5.11 Lip Sync 23
2.5.12 Echo Cancellation 23
2.5.13 Error Control 24
2.5.14 Performance Impairments 25

2.6 Planning a VTC System 25
2.6.1 User Survey 26
2.6.2 Develop a Technical Approach 26
2.6.3 Cost Estimate 26
2.6.4 Schedule 27
2.6.5 Project Proposal 27

Chapter 3 Video Compression Techniques 29
3.1 Overview 29
3.2 A Generic View of Video Compression 31
3.2.1 Coding Techniques 33
3.2.2 Pulse Code Modulation (PCM) 33
3.2.3 Predictive Coding 34
3.2.4 Transform Coding 37
3.2.5 Transformation Techniques 38
3.2.6 Coding of Transform Coefficients 39
3.3 Vector Quantization 45
3.4 Wavelet Coding 47
3.5 Fractal Coding 50
3.6 Object-Based Coding Techniques 52
3.6.1 Generic Unknown Objects 53
3.6.2 Knowledge-Based Coding 55
3.7 Variable Length Coding 55
3.7.1 Comma Code 57
3.7.2 Shift Code 58
3.7.3 B Code 59
3.7.4 Huffman Code 60
3.7.5 Conditional Variable Length Codes 60
3.7.6 Arithmetic Coding 60
3.7.7 Two-Dimensional VLC for Coding Transform Coefficients 61
References 61

Chapter 4 Speech Coding 63
4.1 Speech Coder Attributes 63
4.1.1 Bit Rate 63
4.1.2 Delay 63
4.1.3 Complexity 64
4.1.4 Quality 65
4.1.5 Validation 66
4.2 Currently Available Speech Coders 67
4.2.1 64-56-48 Kbps G.722 SBC 68
4.2.2 16-Kbps G.728 LD-CELP 68
4.2.3 8 Kbps G.729 CS-ACELP 70

Chapter 5 Audiovisual Standards Organizations 73
5.1 International Standards Organizations 73
5.1.1 International Telecommunications Union (ITU) 73
5.1.2 International Standards Organization (ISO) 80
5.1.3 International Multimedia Teleconferencing Consortium (IMTC) 84
5.2 U.S. Standards Organizations 87
5.2.1 Standards Committee T1-Telecommunications 87

Chapter 6 Multimedia Communications via N-ISDN (H.320) 91
6.1 History 91
6.2 H.320: The VTC System Standard 92
6.3 H.261: Video Coding Standard 94
6.3.1 Picture Structure 96
6.3.2 Example of Block Coding 99
6.3.3 Motion Compensation 101
6.4 H.221: Frame Structure for a 64-Kbps to 1920-Kbps Channel in Audiovisual Teleservices 103
6.5 H.242: System for Establishing Communication Between Audiovisual Terminals Using Digital Channels Up to 2 Megabits per Second 104
6.6 H.230: Frame Synchronous Control and Indication Signals for Audiovisual Systems 105
6.7 Audio Coding 105
6.8 Data Channel 106
6.9 Multipoint 107
6.10 Privacy 108
6.11 Narrowband ISDN (N-ISDN) 108
6.11.1 Basic Access 109
6.11.2 Primary Access 109

Chapter 7 Multimedia Communications via the PSTN and Mobile Radio (H.324) 111
7.1 Overview 111
7.2 H.324: Terminal for Low-Bit Rate Multimedia Communication 111
7.3 G.723.1: Speech Coder for Multimedia Telecommunications Transmitting at 5.3/6.3 Kbps 114
7.4 H.263: Video Coding for Low-Bit Rate Communication 115
7.5 H.245: Control Protocol for Multimedia Communications 117

7.6 H.223: Multiplexing Protocol for Low-Bit Rate Multimedia Communication 120
7.6.1 Multiplex Layer 120
7.6.2 Adaptation Layer 123
7.7 Data Channel 123
7.8 Audiovisual Telephony via Mobile Radio 125

Chapter 8 Multimedia Transmission via B-ISDN (H.321, H.310) and LAN Networks (H.322, H.323) 129
8.1 Broadband ISDN 129
8.1.1 Constant Bit Rate (CBR) and Variable Bit Rate (VBR) Coding 133
8.2 Adaptation of H.320 Visual Telephone Terminals to B-ISDN Environments (H.321) 133
8.3 High-Resolution Broadband Audiovisual Communication Systems (H.310) 134
8.3.1 Communication Mode 134
8.4 Visual Telephone Systems for Local Area Networks That Provide a Guaranteed Quality of Service (H.322) 138
8.5 Visual Telephone System for LANs That Provide a Nonguaranteed Quality of Service (H.323) 139
8.5.1 Scope 139
8.5.2 Terminal Characteristics 142
8.5.3 Definitions 146

Chapter 9 Multipoint Graphic Communications (T.120) 149
9.1 Node Controller 150
9.2 Communications Infrastructure 150
9.2.1 T.122/125 Multipoint Communications Service (MCS) 152
9.2.2 T.124 Generic Conference Control (GCC) 154
9.2.3 T.123 Transport Protocol Stack Profiles 155
9.3 Application Protocols 157
9.3.1 Recommendation T.126: Still Image Exchange and Annotation (SI) 157
9.3.2 Recommendation T.127: Multipoint Binary File Transfer Protocol 158

Chapter 10 ISO Audiovisual Standards (MPEG, JPEG, JBIG) 159
10.1 MPEG1 160
10.1.1 System Layer (11172-1) 160
10.1.2 Video Coding (11172-2) 160
10.1.3 Audio Coding (11172-3) 161

10.2 MPEG2 (13818) 161
10.2.1 MPEG2 System (13818-1) 161
10.2.2 MPEG2 Video (13818-2) 162
10.3 MPEG4 (14496) 166
10.4 JPEG Coding Algorithm 168
10.4.1 The Baseline System 169
10.4.2 Extended System 171
10.4.3 Lossless Coding 172
10.5 The JBIG Coding Algorithm 172

Glossary 175

Acronyms and Abbreviations 187

About the Author 189

Index 191

Acknowledgments

This book frequently references telecommunication standards that have been, and are being, developed by the ITU and ISO standards organizations. Consequently, I wish to acknowledge the work of all the experts and organizations that have contributed to these standards over the years. I also wish to recognize a few particular individuals for their unusual contributions: Sakae Okubo for his chairmanship of several key ITU expert groups; Richard Cox, who was the key contributor to the chapter on speech; Gary Thom for his work on the H.323 Standard; and Neil Randall for his many contributions over the years.

Introduction 1

All of our lives have been revolutionized by three developments from the world of electronic communications, namely, the telephone, the television, and the computer. The telephone has been a pervasive fundamental communications tool in both home and business environments for many decades. Although life at home has been turned upside-down by broadcast television for many years, TV is just beginning to seriously impact the business world. The computer is the most recent explosion in our daily lives, giving rise to vast quantities of *data* to be shared. These three signals—audio, video, and data—are now being merged under the new banner of multimedia.

This book describes recent developments in the technology and standards for the communication of multimedia signals. Examples of multimedia communications include the videophone, video teleconferencing (in the conference room as well as on the desktop), and remote access to multimedia databases. Although much of the technology and many of the standards covered in this book are applicable to the world of broadcast entertainment TV, this is not the focus. Instead, this book is concerned with person-to-person, or person-to-database, interactive multimedia communications. Communication channels utilized to carry these multimedia signals include the telephone network, the *integrated services digital network* (ISDN), LANs, and mobile networks.

The markets for video teleconferencing and the videophone are exploding. There are three major reasons for this breakout, improved audiovisual quality, reduced cost, and communication standards. Recent breakthroughs in video and audio compression technology are responsible for the improvement in quality. An overview of this video and audio compression technology is provided in Chapters 3 and 4, respectively. The cost of *video teleconferencing and videophone* (VTC/VP) systems has been drastically reduced in two fundamental areas, namely, communications and the VTC/VP terminal itself. The communication cost has been radically reduced because the transmission bit rate has dropped sharply, and the cost/bit from the common carrier has also been cut.

VTC/VP terminal cost has been drastically reduced because of the incredible strides in integrated circuit development. Last, but not least, the VTC/VP revolution could not have occurred without the development of communication standards. A few years ago the VTC/VP market was frozen because different vendors' terminals could not talk to each other. They each used different proprietary coding algorithms. In 1990 the *International Telecommunications Union* (ITU) finalized the H.320 VTC/VP standard, and today every manufacturer provides the standard in its codec. The standard accomplishes a number of objectives. It assures interoperability between terminals manufactured by different vendors. It also reduces terminal cost because chip manufacturers produce devices implementing the standard in high volume—therefore, low cost.

The market for *video teleconferencing* (VTC) and videophone systems has several dimensions. The VTC market in the business community is relatively mature and is growing rapidly. In this application the transmission bit rates range from 128 kbps for lower level elements of an organization up to 384 kbps for management levels. Fixed and roll-around units are installed in conference rooms, while systems using the personal computer platform are proliferating on the desktop. Growth in the desktop area is being spurred by the recent acceptance, and recognized productivity, of telecommuting. Workers are using their personal computers at home as platforms for desktop videoconferencing back to work and with the world at large. In many of these applications, it is recognized that the *data* element of the multimedia communications (for example, white board, spread sheet, and document editing) is extremely important, perhaps more important than motion video.

The market for the videophone is not as mature as VTC due to the marginal quality that has been available from recent products and the high cost. This will change rapidly because of the quality improvement inherent in H.324 and H.320 standards as well as the price reduction that will result from mass production of standardized products. The videophone will penetrate the business market first but, more importantly, will be rapidly adopted for use in the home by the consumer. The two major applications will be the person-to-person videophone and database interactivity (for example, shopping, banking, reservations, and consulting). It is likely that the most important use will be the nonconversational database application.

The major part of this book is concerned with the new standards that define the procedures for multimedia communications. Figures 1.1 and 1.2 illustrate the key ITU recommendations that have been, and are being, developed for VTC/VP. H.320, designed for operation over the N-ISDN, is the most mature standard (established in 1990) and forms the cornerstone of all videoconferencing systems—room-based as well as desktop. H.320 is available from all vendors and guarantees interoperability between systems from different manufacturers.

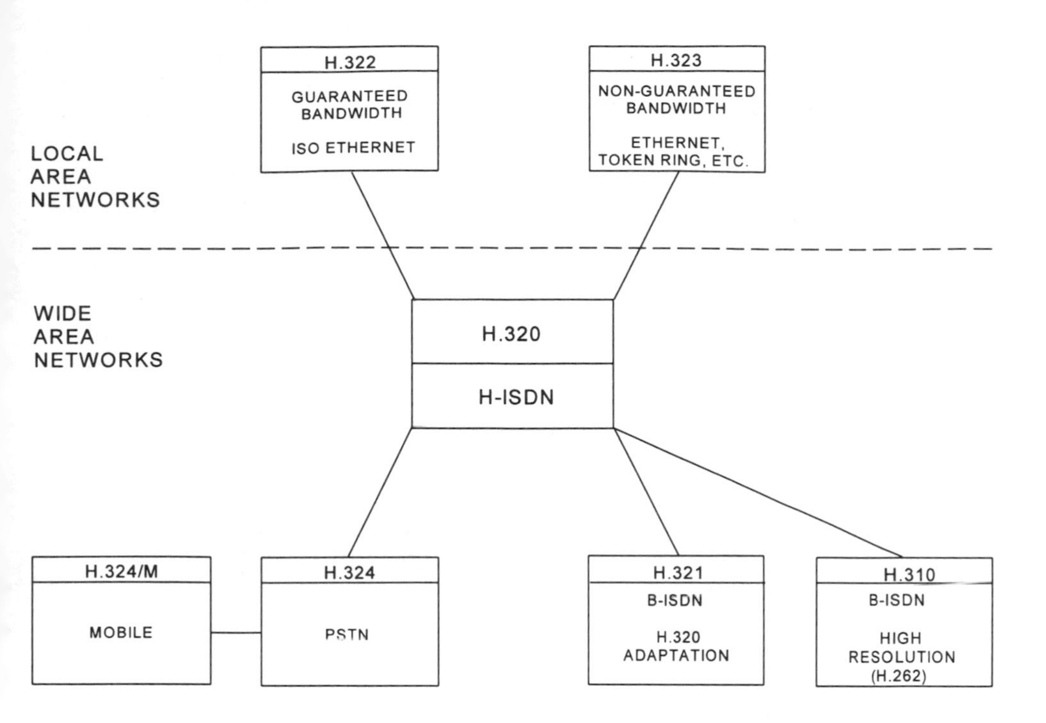

Figure 1.1 Multimedia communication standards.

NETWORK		TERMINAL STANDARD					
		Overall	Video	Mux	Control/ Signaling	Audio	Comm. Interface
WAN	PSTN	H.324	H.261/H.263	H.223	H.245	G.723.1	V.34 MODEM
	MOBILE RADIO	H.324 (C)	H.261/H.263	H.223 (A)	H.245	G.723.1 (C)	MOBILE RADIO
	N-ISDN	H.320	H.261	H.221	H.242	G.711 (M) G.722 (O) G.728 (O)	I.400
	B-ISDN/ATM	H.321	H.261	H.221	Q.2931	G.711 (M) G.722 (O) G.728 (O)	AAL I.363 AJM I.361 PHY I.400
		H.310	H.261/ H.262*	H.222.0*/ H.222.1	H.245	G.711 (M) G.722 (O) G.728 (O) MPEG 1,2	AAL I.363 AJM I.361 PHY 1.432
LAN	ISO ETHERNET	H.322	H.261	H.221	H.242	G.711 (M) G.722 (O) G.728 (O)	I.400 & TCP IP
	ETHERNET, TOKEN RING	H.323	H.261	H.225.0	H.245	G.711 (M) G.722 (O) G.728 (O)	TCP IP

* common text with ISO MPEG 2

Figure 1.2 ITU audiovisual recommendations.

Recommendations H.324, H.321, H.310, and H.322 were "frozen" by the ITU in February 1995 achieved final approval in 1996. H.324 defines a multimedia terminal that communicates speech, data, and video signals over the public-switched telephone network. As part of the H.324 terminal, the new standards established for speech coding, video coding, control, and multiplex are G.723.1, H.263, H.245, and H.223, respectively.

H.321 and H.310 are new recommendations defining videoconferencing terminals for transmission via the B-ISDN/ATM network. H.321 merely converts the H.320 standard from N-ISDN operation to B-ISDN transmission; all of the H.320 infrastructure (H.261, H.221, H.242) remain intact to maximize interoperability between the two networks. H.310 is a new recommendation that adapts the ISO MPEG-2 standards for communication over the B-ISDN/ATM network. H.262 and H.222.0 are common text standards with ISO MPEG-2.

H.322 adapts the H.320 standard for those LAN networks that guarantee the bandwidth (for example, ISO Ethernet). H.323 is a new standard designed to provide videoconferencing over non-guaranteed-bandwidth LANs such as Ethernet and Token Ring. The H.323 standard has been frozen and will be fully approved in 1996.

The reader should note the H.261 video coding standard is mandatory for all the recommendations listed in Figure 1.1. This greatly enhances interoperability between networks. A great deal of commonality also exists in the audio, multiplex, and control standards.

All the aforementioned recommendations, including those listed in Figure 1.1, define multimedia terminals that provide for the transmission of audio, data, and video signals. In many applications, particularly those for the desktop, data originating on the PC/workstation platform is the most important information to be transmitted. The ITU is developing the T.120 series of recommendations for the communication of this data. T.120 defines a protocol stack that can be used for data-only transmission; in addition, the T.120 protocol is used for the data component in the multimedia terminals such as H.320, H.324, H.321, and H.322.

Video Teleconferencing: Benefits and System Design

2

2.1 INTRODUCTION

A teleconference is a meeting implemented between people who are physically separated from each other, achieved by using electronic communication techniques. Teleconferencing systems can be broadly classified into three categories: audio-only, audiographic, and full-motion video. Of course, a full-motion video system includes audio and can include graphics as an option.

As a further classification, there are two types of full-motion video teleconferencing systems, interactive and broadcast business television. In the case of broadcast business television, the typical application is a single event (for example, a product announcement) with one transmitter and many receivers. The primary focus of this chapter is interactive videoconferencing as opposed to one-time broadcast events. Interactive videoconferencing includes groups of people in a conference room or a single person using a desktop videophone.

2.2 BENEFITS OF VIDEO TELECONFERENCING

The benefits that organizations can gain from videoconferencing are outlined as follows.

- *Faster decision making.* Since a video conference can be established with little more difficulty than an in-house meeting, people separated by miles can come together and share ideas and information when the need arises. There is no need to delay a decision until all participants can clear days for travel on their calendars. In industry, this is particularly valuable in rapidly resolving unexpected crises.
- *Better decisions.* Because there is no additional cost for additional meeting participants, the conference can include all the people that are involved in the undertaking. If travel were involved, this might not be possible,

even if the organization were willing to incur the cost. Also, information sources not originally considered may be brought into the meeting. People may join a video conference as easily as they can an in-house meeting.

- *Increased productivity.* Valuable employees need not waste time traveling to and from airports, waiting in check-in lines, and sitting on the tarmac awaiting takeoff clearance. If a meeting takes two hours, the participants will expend only two hours of their valuable time.
- *More meetings.* VTC permits an organization to have more meetings than otherwise would be feasible. More meetings to manage a project, for example, could result in a higher quality result or more rapid completion.
- *Increased employee safety.* Travel involves some risk. A video conference eliminates this concern.
- *Tighter security.* If coworkers travel together and discuss business, information even though unclassified, may be compromised if their discussions are overheard.
- *Improved employee morale.* Some travel is enjoyable. However, frequent trips, particularly to remote locations or second-tier cities, can be depressing to employees. Videoconferencing cannot eliminate the need for travel, but it can reduce travel requirements significantly.
- *Avoided travel costs.* This is the most easily quantified benefit. Many costs are associated with official trips. The traveler must pay for conveyance to the airport or pay for parking. At the destination city, there may be the need for a rental car or a taxi. Some trips require an overnight stay in a hotel. Nearly all trips call for payment of meal costs. And, of course, the cost of an airline ticket or alternate transportation must be included. Employees can seldom take advantage of discount rates that require advance payment or weekend stays.
- *Reduced fatigue.* Business travel is frequently a tiring, frustrating experience that can result in fatigue and consequential poor performance during the meeting as well as after the traveler has returned home. In the case of long-distance travel, it can take several days for this fatigue, or jet lag, to diminish. Upon occasion, the employee is truly sickened by the trip. Obviously, the elimination of this fatigue is an important VTC benefit that is difficult to quantify.
- *Efficient use of key personnel.* One of the VTC benefits that is difficult to quantify is the ability to efficiently employ and apply critical personnel. For example, there are cases in industry where it would be desirable for a particular key person to attend a meeting in Europe in the morning and in Japan in the afternoon. In effect, VTC makes this feasible.
- *More disciplined, productive meetings.* In many cases, more is accomplished at a VTC meeting than at a face-to-face meeting because all participants realize that it is necessary to operate in a more disciplined manner

at the VTC meeting. There are fewer interruptions, and there is more listening. One reason for this discipline is the fact that most VTC meetings have a fixed time duration.

- *Team building.* When meetings are held at the normal frequency, it is typically necessary for a project to be organized on a basis where all decisions and control originate at the top and are fed down throughout the organizational structure. This is often necessary to meet program schedule requirements. Since VTC permits more frequent meetings, it becomes possible to more frequently reach decisions by consensus rather than top down, thereby achieving a sense of a team effort, which in turn results in a greater degree of motivation and energy by all members of the organization.
- *Interning.* It is frequently convenient to include nonparticipatory observers in a VTC because it can be done with very little cost relative to face-to-face meetings. This provides an inexpensive way to train personnel by exposing them to a particular project and management procedures.
- *Reliability.* In many cases, a face-to-face meeting is canceled, or its effectiveness reduced, because of terminated flights or bad weather, for example. VTC eliminates this problem.

2.3 VIDEO TELECONFERENCING OVERVIEW

At the highest level, a VTC system consists of a number of VTC terminals, or nodes, interconnected by means of a communication network. Such a configuration is illustrated in Figure 2.1. Full duplex communication channels are used to create a real-time interactive environment (for example, motion video, audio, and graphics) between remote sites. The VTC terminals can vary greatly in complexity, ranging from a large multiroom configuration down to a desktop videophone device. In general, the transmission bit rate varies with the complexity. Large complex rooms used by many people and executive level personnel employ high transmission bit rates (for example, 1.544 Mbps and 768 Kbps). Intermediate-sized rooms used typically for project work will employ intermediate bit rates (for example, 384 Kbps or 128 Kbps). Desktop videophones will employ the lowest rate (for example, 64 Kbps and 128 Kbps).

Although point-to-point VTC connections are very important, multipoint connections have been found to be even more important. Multipoint permits a large number of terminal nodes to simultaneously participate in a conference. Multipoint connections are accomplished by each terminal connecting to a *multipoint control unit* (MCU).

VTC systems are conveniently classified into three categories as defined by their physical configuration, namely, customized conference room, rollabout module, and desktop. Table 2.1 summarizes the key characteristics of these three different types of systems.

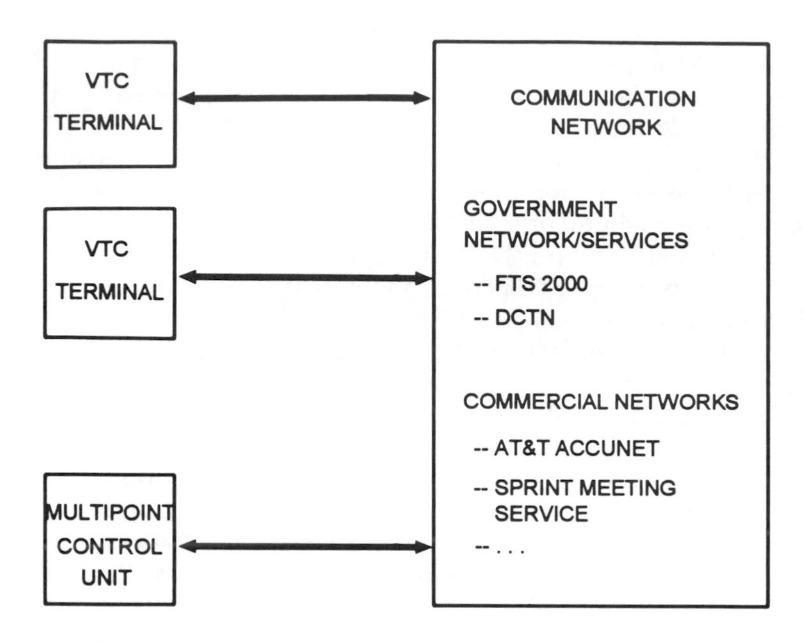

Figure 2.1 A generic multinode VTC system.

In general, if the VTC requirement calls for a number of people at a facility to routinely take part in remote conferences, there is a need for a customized room or a room with a rollabout to accommodate people and to provide a convenient mechanism for integrating graphics and documents into the conference. If there is a requirement for a large number of people in the room (more than six) and if the personnel taking part in the conference are very senior, it may be necessary to implement a large customized conference room. If, however, the typical number of people using the room will be six or less, it is usually possible to place a rollabout modular unit in an existing room to minimize installation cost.

The key ITU standards and their function in a VTC terminal are as follows.

- H.320: Narrow-band visual telephone system and terminal equipment;
- H.261: Video codec for audiovisual services at P × 64 Kbps;
- H.221: Frame structure for a 64-Kbps to 1920-Kbps channel in audiovisual teleservices;
- H.242: System for establishing communication between audiovisual terminals using digital channels up to 2 Mbps;
- H.230: Frame synchronous control and indication signals for audiovisual systems;

Table 2.1
Physical Configuration of VTC Systems

VTC Configuration	*Physical Layout*	*Typical Display*	*TV Camera(s)*	*Lighting Acoustics*	*Typical Scene: Number of People*	*Microphone/ Speaker*
Customized room	Large conference room, large table; possible additional chairs. Electronic equipment is in a back room.	Rear screen projector(s) or multiple large TV monitors built into a wall.	Usually multiple cameras with pan, tilt, zoom.	Usually customized.	Large group of people; can be more than six.	Multiple microphones or one table top unit.
Rollabout	Self-contained module(s), table within a conference room.	One or two large TV monitors built into rollabout module; small TV window for self-view.	One or two cameras with pan, tilt, zoom.	Usually normal room lighting and acoustics.	Small group of people—up to six.	Typically one table-top unit.
Desktop	Camera/monitor on desktop; electronics in videophone or on floor.	One small TV monitor.	One small camera. No pan, tilt, zoom.	Usually normal room lighting and acoustics.	Talking head; head and shoulders; usually one person.	Handset or integrated into videophone.

- G.711: *Pulse code modulation* (PCM) of voice frequencies;
- G.722: 7-kHz audio-coding within 64 Kbps;
- G.728: Coding of speech at 16 Kbps using *low-delay code excited linear prediction* (LD-CELP);
- H.231: Multipoint control unit for audiovisual systems using digital channels up to 2 Mbps;
- H.243: System for establishing communication between three or more audiovisual terminals using digital channels up to 2 Mbps.

2.4 VIDEO TELECONFERENCING ROOM CONSIDERATIONS

2.4.1 Room Size, Furniture, and Location

The room designer must consider the types of meetings that will take place in the organization, such as small working group sessions, larger information-sharing meetings, or very large conferences, and select the most sensible video conference room design for the organization. Before searching for a location for a video conference room, it is necessary to know how much space is required, which depends on the number of people to be accommodated and the type of equipment to be used.

Let us begin with the number of on-camera participants. Each person should be allowed 3 feet of space along one side of the conference table. Thus, if the room is designed for six active conferees, an 18-ft-long conference table should be provided. In addition, there will need to be at least one aisle at one end of the conference table, which adds another 2.5 feet per aisle to the width of the room. We are now at 20.5 ft to 23 ft.

If a graphics camera is required, space must be allocated. If a separate cabinet is envisioned, another 2.5 ft by 2.5 ft of floor space will be required. Since many users prefer to have this capability near the conference table, the designer should consider allowing space for the unit in calculating the width of the room. Of course, ceiling-mounted graphics cameras do not require additional floor space.

There may be a requirement for additional off-camera seating. If the space under consideration is shaped to accommodate this area along the side, it will require another 2.5 ft per row of chairs and 2 ft per aisle between rows. Therefore, we have a room of at least 21.5 ft—ideally 25.5 ft or more—in width.

In determining the depth of the room, we must begin with the distance from the camera to the center of the conference table, which is typically 10 ft. The seats will take up another 3 ft when occupied, and an aisle of 3 ft should be provided. Therefore, the minimum distance from the camera to the rear wall is 16.5 ft. Some designs allow for extra seating behind the active participants rather than off to the side. A layout of a typical customized conference room is shown in Figure 2.2.

If a rollabout cabinet will house the room equipment, another 2 ft must be provided to account for the depth of the cabinet. There should be 6 in of clearance from the rear wall to facilitate air flow. If the installation sets the equipment into the front wall, there will need to be an equipment room to allow access to the equipment; allow at least 6 ft for this purpose. Therefore, a room with a cabinet system will need to be approximately 19-ft long, and a

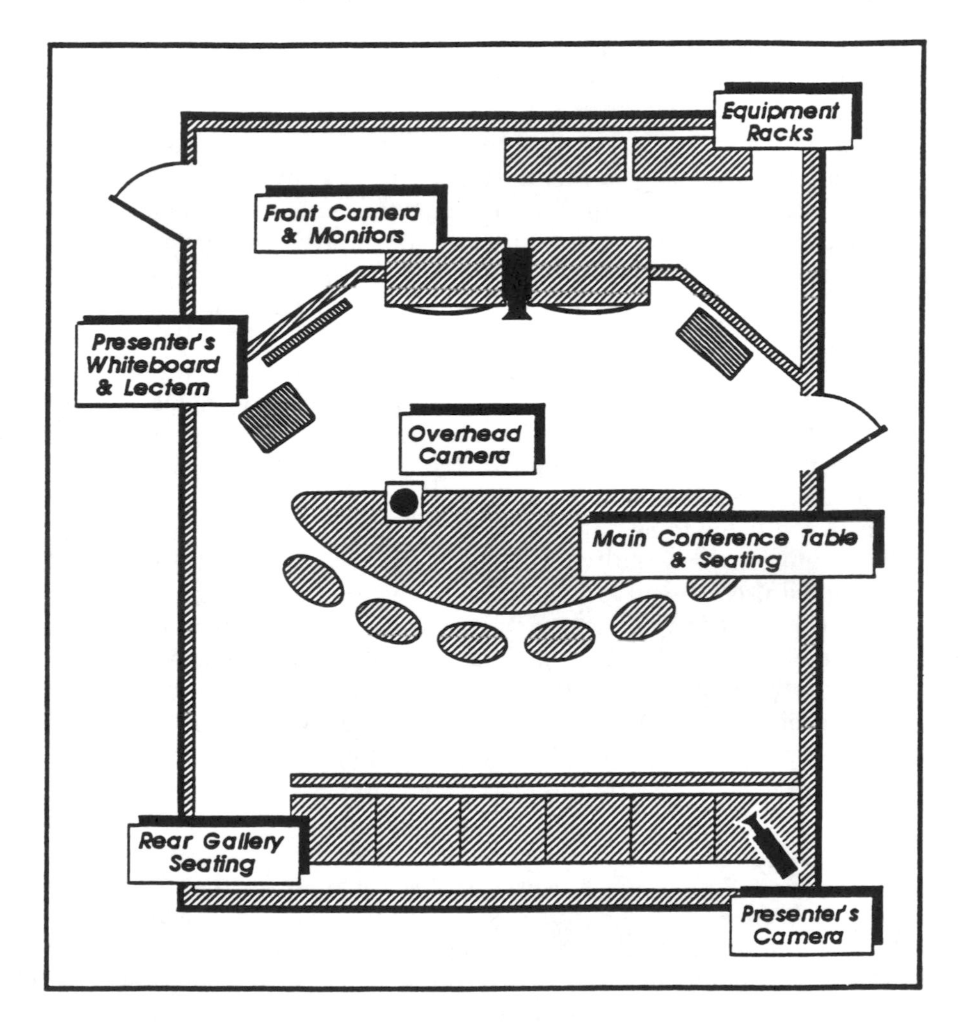

Figure 2.2 Layout of a typical customized conference room.

site-built facility that accommodates the same number of participants will need to be 22.5 ft. Of course, it is necessary to consider doors in the room layout, that is, doors for access to both the conference room and rear equipment room.

It is desirable to keep the ambient noise level of the room at a low level (45 dB or less). This is achieved by a combination of site selection and/or sound treatment of the walls. Room reverberation time should be between 0.3s and 0.5s. The room's absorption coefficient should lie between 0.25 and 0.45. If less than 0.25, the room will be hollow sounding; if more than 0.45, it will sound dead.

Most installations have an oval table with on-camera participants seated on one side of the table. Usually, the arc of the oval is designed so that every participant is equidistant from the monitor and camera. Everyone can see equally as well, and the only size differences apparent to the people viewing a distant monitor are those that actually exist. In this configuration, the people at another location are electronically on the other side of the table. This "we against them" seating arrangement may create an artificial barrier to effective group dynamics. If this is a major concern, another table shape should be considered.

A trapezoidal or blunted triangular table can be used to allow for two or three people on each of the two equal sides, creating the illusion of everyone seated at a square table, with four equal sides and no clear leadership position. A drawback is that those seated closest to the camera will appear larger than those farther away from the camera.

The chairs selected should not have visually reflective parts that would cause the room lights to be reflected directly into the camera.

It is important to carefully locate the conference room within the facility. If the intended users are all top executives, it would be desirable to locate the room near the executive office area. If a mixture of senior level managers and other employees are expected to use the installation, it would be better to locate it in a neutral location. Try to avoid putting the video conference room in an obscure part of the building. If possible, the location should not be adjacent to heavily traveled, noisy corridors. Outside walls and space adjacent to elevators shafts or equipment rooms should also be avoided.

2.4.2 Cameras

Generally, a camera can frame three seated participants and capture enough detail to communicate facial expressions or subtle body motions. For this reason, a typical room design with a six-person table requires two cameras, each viewing three people.

Since all VTC systems transmit only one video signal, means must be provided to integrate the two camera signals into the one video transmission channel. This is accomplished by either switching between the two cameras or using a split-screen technique. Camera switching can be initiated either manually or by voice activation. Voice activation is somewhat preferred because the picture of the speaker is automatically transmitted. Manual switching can be greatly simplified by using "presets" where a number of specific camera orientations (for example, pan, tilt, zoom, and focus) are stored in the system. A single operator command causes the camera to move quickly to a preset location.

Until now, we have only discussed cameras that view the participants, that is, "people cameras." In most VTC facilities, there is also a requirement to handle some visual aids such as flip charts, slides, transparencies, or solid objects. A separate camera is generally provided for this requirement. In some systems, this document camera is mounted in the ceiling or on a stand that can be placed on the conference table.

In rollabout systems, a separate graphics cabinet is often used. This cabinet may house a camera, a slide projector, a light table for transparencies, and a mounted lighting surface for capturing documents or small objects.

The ability to focus a camera on either a white board or a podium is required in some installations. A single camera with a present view or a separate camera is typically used for this requirement.

2.4.3 Display System

Early video conference room designs used video projection units instead of monitors because large monitors were not available. There are several problems with projection systems, not the least of which is the high cost. The image is not sharp and clear due to the enlargement. The image is duller than monitor displays and must be viewed straight-on to avoid a further reduction in brightness. Finally, the camera is far from the center of the screen. In summary, video projectors are rarely used in VTC facilities; they are used only where there is a very large number of participants.

Today, monitors as large as 35 in (diagonal dimension) produce bright, clear images. The correct monitor size is determined by the distance between the monitor and the main participants. The general rule is that the height of the monitor should be one-eighth the distance between the screen and the viewer. In other words, if the participants are seated 10.5 ft from the monitor, the monitor height should be 15.75 in and the diagonal size of the monitor should be 26.25 in. As this is not a standard monitor size, monitors of either 25 in, 27 in, or 30 in may be used with confidence. A 35-in monitor would be the upper limit for a room this size. The center of the monitor screen should be approximately 3.5 ft above floor level.

Many VTC display systems being installed today used two TV monitors—one for "people pictures" and one for graphics. The scene viewed by the local camera is usually presented as a picture-in-picture (small window) on the "people picture" monitor. It is also common to implement a VTC system using a single monitor particularly with a fully packaged rollabout as shown in Figure 2.3. In such a system, it is obviously necessary to switch the function of the monitor between people picture and graphics or to use a window for the people pictures. Again, the local camera screen would be viewed using a small window.

2.4.4 Audio

Audio is the fundamental cornerstone of any teleconference system; graphics or video, although very important in most cases, must be considered to be

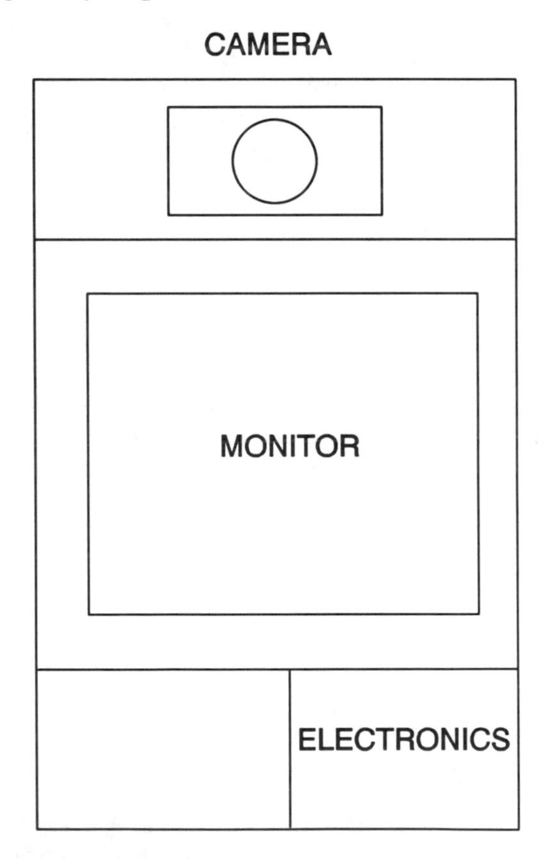

Figure 2.3 Front view of a rollabout unit with one monitor.

supplementary. Unfortunately, there are many characteristics of an audio teleconference system that are very different from a conventional telephone system. Consequently, much of the telephony technology that has been developed over the decades is not applicable to the audio teleconferencing problem. Technical challenges that are unique to audio teleconferencing include room acoustics, microphone/speaker placement, echo cancellation, and bridging for multipoint operation.

The acoustic characteristic of the teleconference room can significantly affect the quality of the audio system. Ambient noise, acoustic reflectivity of the walls, and reverberation time are major factors contributing to audio quality. The technical characteristics and placement of the microphone(s) and speaker(s) in the room are also important factors in the performance of the audio system. The factors will determine the tendency of a system to "howl" from feedback or generate an echo that must be canceled or suppressed.

Many rooms contain multiple microphones distributed around a conference table. The audio from these microphones can be combined in three different ways to form one signal for transmission. They can be mixed, manually switched, or automatically switched according to voice activity.

An objectionable echo can occur when voice signals are sent back to the originating end as a result of speaker-to-microphone coupling. The echo problem is solved either by suppression or cancellation. Echo suppressers automatically reduce the level of the return audio path when a local participant is talking, thereby reducing or eliminating the associated echo. Echo cancelers generate a pseudo-echo based upon the incoming audio signal and subtract this pseudo-echo from the transmitted signal containing the echo. In general, echo cancelers are more complex than echo suppressers and typically provide higher voice quality. Echo suppressers can exhibit a voice clipping characteristic that in some cases can be disturbing.

In the case of multipoint teleconferencing, an audio bridge is used to mix the voice signals originating from the several conference locations and distribute the output mixed audio to the conferees. The audio bridge must also control the magnitude of the output composite signal to avoid system overload.

In summary, the audio portion of a teleconference is absolutely crucial to the success of the conference. It is a technical challenge to design a system that provides high-quality audio under all conference circumstances.

In recent months, an integrated audio product has appeared on the market from several vendors that shows promise of significantly easing the VTC audio problem. The device integrates microphone, speaker, and echo canceler into a single device that sits in the center of the table. This unit has promise for minimizing the need to acoustically treat the room and to be concerned with placement of microphone and speaker.

2.4.5 The VTC Controller

The remaining piece of equipment to discuss is the VTC controller. It is the central nervous system of the installation. The user controls and interacts with the equipment via the control unit. The control system must be user friendly. It should have the capacity to be expanded to meet the evolving needs of the organization, and of course, it must be reliable.

Control units can be divided into the two broad categories of wired or infrared.

2.4.5.1 Wired Control Units

Typically, wired units are a custom-designed array of mechanical controls such as buttons, switches, sliding potentiometers, and joy sticks. The front panel is custom inscribed to indicate control functions. Another popular design uses a touch-sensitive screen. Here again, the display can be customized to make the controller's operation easily understood by users.

All of these control units require a cable to connect them to the equipment room or rollabout cabinet. For this reason, wired control units are difficult to move around the conference table. The stiff cables and bulky units do not lend themselves to shared control of the meeting. In fact, some designs have the control unit built into the table, which is a limitation in an organization that encourages participation among meeting attendees. The person or people seated adjacent to the control panel become the meeting leaders.

If a podium or white board is included in the installation, some of the controls will need to be duplicated and installed near the presenter. Also, the cable can be a physical hazard and must be installed in a duct. These wiring requirements add to the system's cost, and in the case of a rollabout, the transportability is reduced.

On the plus side, there is no space limitation on the face plate of a wired control panel. The designer may make the controls easy to use and very explicit. The type of control favored by the majority of users may be installed. For example, a slide control may be favored for the volume, while a joy stick may be easier to use when controlling the cameras. Other controls may be added to an existing wired controller as long as spare cable capacity was provided initially. Figure 2.4 illustrates the layout of a typical room controller.

2.4.5.2 Infrared Control Units

Infrared units may be either hand-held or table-top devices. Therefore, the number of control functions is limited. System engineers often get around this

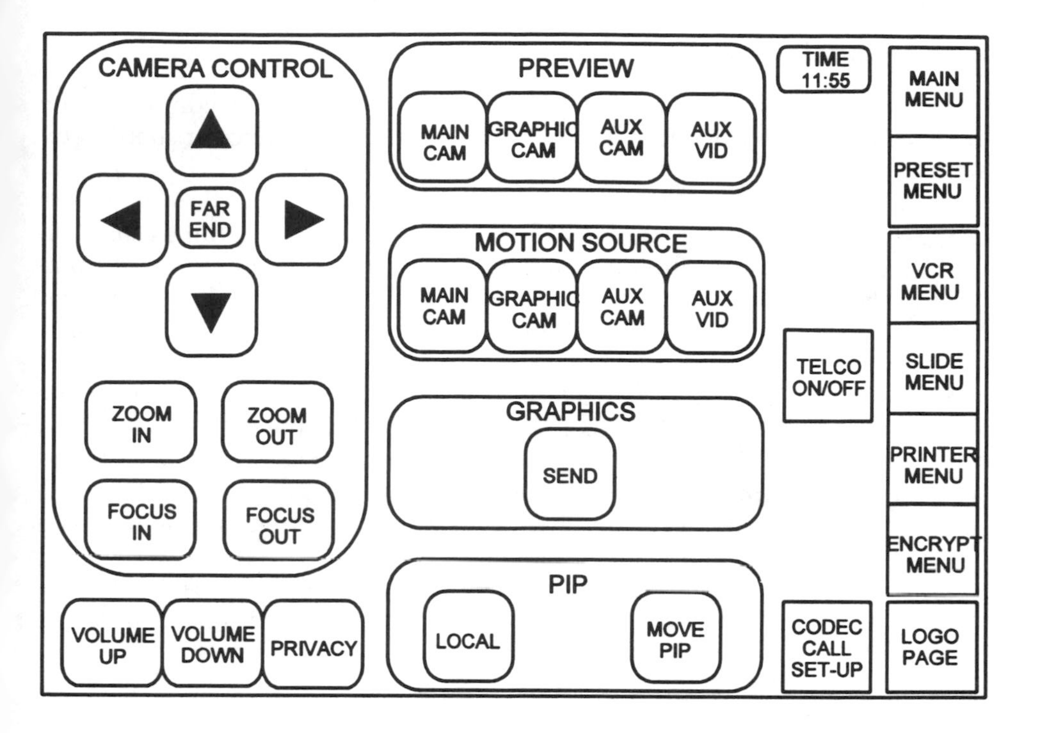

Figure 2.4 Layout of typical VTC controller.

constraint by giving some buttons multiple uses. The limited space on the face of the controller also restricts the amount of information that can be inscribed on the face plate. Users definitely require training to use these controllers. On a hand-held unit, only buttons are available. Other options such as slides or joy sticks cannot be easily implemented.

Table-top infrared units can utilize more options. However, they are limited to a finite number of codes, with a separate code required for each function. Space is not a restriction. The face plate may be as explicit as the designer deems necessary.

There are also problems of lesser concern with this approach. All infrared units require batteries. These can begin to malfunction during a meeting and cause a great deal of user frustration.

The major strength of an infrared design is the ease with which the control of the meeting can be shifted. Not only can the unit be passed around, but there may be more than one unit in the room. For example, a separate controller can be placed at the location of the presenter and used when appropriate.

2.4.6 Administration

It is necessary to allocate significant personnel resources for the administration of a VTC room. The work activities of a VTC room administrator or coordinator are as follows.

- *Scheduling of conference calls*: This work can become complex, particularly to schedule multipoint conferences for a busy conference room. Computerized systems are available today to assist in the scheduling process.
- *Training*: There is a continuous need to train users, operators, and maintenance personnel.
- *Promotions*: It has been clearly established that a conference room is more productive if the use of the room is promoted within the user's community.
- *Operation*: Some complex VTC room systems require the use of a dedicated operator to support the users.
- *Maintenance*: The room coordinator must arrange for periodic and on-call maintenance of the room equipment.
- *Security*: Some conference rooms must handle encrypted calls. In these cases, considerable administrative effort is required to support the encryption equipment and related data.

2.5 QUALITY AND PERFORMANCE CONSIDERATIONS

2.5.1 General

"A picture is worth a thousand words." This old saying not only gives the justification for teleconferencing but also states the fact that a picture contains much more information than a spoken word. Transmission of such information between distant locations requires an *electronic highway* that is wide enough to handle all required information. Greater width translates into higher cost. Therefore, in the past, pictorial transmission could be economically justified only for high-value applications such as TV broadcasting. Even though recent developments in transmission technology have considerably reduced this cost factor, it is likely to remain important for many years to come. Video teleconferencing has become economically viable only by using special techniques that drastically reduce the amount of required transmitted information.

Fortunately, every picture contains a large amount of redundancy. There are always areas that do not contain details and, thus, do not call for the transmission of a large amount of information. Even with a large amount of detail, if the detail does not change rapidly, very little information must be transmitted. Furthermore, even in the case of a fast moving scene, the changes at any one point are limited and often predictable. Thus, it is not always necessary to transmit a complete picture. Limiting the transmitted information to changes is generally sufficient to reconstruct the picture at the receiving location. Applying such techniques can result in a large amount of bandwidth compression or data rate reduction.

Unfortunately, everything has its price. Most compression techniques result in at least some loss of picture quality. The development of more advanced techniques makes this loss less and less noticeable. Furthermore, a teleconference normally does not call for the high picture quality that we are accustomed to in entertainment TV broadcasting. This makes the application of compression techniques both technically and economically viable.

2.5.2 System Considerations

Video teleconferencing is still a relatively new technology. Many manufacturers in the United States, Japan, and Europe have developed techniques for video compression. In the past, most of these techniques were proprietary, meaning that equipment of different manufacturers could not operate with each other. In the beginning, this was no problem since teleconferencing systems were small and independent of each other. However, after a few years of development, it became obvious that this lack of interoperability presented an intolerable constraint, not only within the United States but even more so in global applications where different TV standards (such as NTSC, PAL, and SECAM) have to be accommodated. Therefore, the ITU developed a standardized technique (algorithm) for digital video encoding, Recommendation H.261, which can be made compatible with all national TV broadcast standards. Of course, equipment with proprietary algorithms will be in use for some time to come, but for future systems only equipment conforming with H.261 and its associated standards need and should be considered.

2.5.3 Picture Format

None of the existing analog TV broadcast standards is suitable for compressed video transmission. Compression techniques call for a digital signal, which requires breaking down the picture into individual samples or elements (pixels). Most important information is contained in the monochrome or luminance portion of the picture; therefore, transmission is performed in separate luminance and chrominance (or color difference) components at different levels of quality. Standard sync and blanking signals can be regenerated in the receiver and need not be transmitted, thus using all the available transmission channel for picture information.

2.5.4 Common Intermediate Format

The H.261 standard defines a *Common Intermediate Format* (CIF) as seen in Table 2.2. The best obtainable resolution is roughly half that of the NTSC broadcast standard, which is roughly comparable to the resolution provided by a consumer-quality VCR. This resolution is suitable for group conferences, where each participant takes up only a small fraction of the viewing area. However, if the available bit rate is low, the result may be jerky motion. For graphics, CIF can be used to transmit simple viewgraph text images that have about 15 or fewer lines of text. Note also that the CIF is not required of all codecs, so conferences with other codec models may not be able to operate at CIF resolution. CIF is recommended for all codecs that will be operated at bit rates of 384 Kbps and above.

2.5.5 QCIF

The lower quality *Quarter Common Intermediate Format* (QCIF) is limited to only half the resolution of CIF in each dimension and, therefore, has one-quarter the number of pels as CIF. Clearly this resolution is noticeably poorer than entertainment quality TV. This resolution is suitable for single-person conferences, where only the head and shoulders of one person are shown. For graphics, QCIF can be used only for very simple viewgraphs that have at most seven lines of text. All H.261 codecs are required to have QCIF, even if they have CIF capability, so it provides a fall-back mode that will work between all types of codecs. Even if a codec has CIF capability, most codecs provide a feature that forces operation at QCIF if it is desired to improve frame rate or to economize on bit rate.

Table 2.2
Common Intermediate Format (CIF)

	Luminance	*Chrominance*	
Parameter	*Y*	*R-Y*	*B-Y*
Full CIF			
Pels/Line	352	176	176
Lines/Picture	288	144	144
Quarter CIF			
Pels/Line	176	88	88
Lines/Picture	144	72	72
Picture Rate: Submultiple of 29.97 Hz			

2.5.6 Frame Rate

The TV broadcast standard of 30 frames per second produces a perfectly smooth picture. This is important for entertainment but not for teleconferencing. Therefore, a very effective way to reduce the transmitted bit rate is not to transmit all frames and "fill in" at the receiver by simply displaying the last transmitted frame several times. The resulting received picture may not show smooth motion. Indeed, it may appear as a series of snapshots. However, this is adequate for many applications. Indeed, many observers find it preferable to a blurry picture with smoother motion. The picture rate, though theoretically allowed to be at the NTSC standard, 30 frames per second, will normally be lower. This feature provides a wide range of freedom in reducing the transmission requirements. A value of 15 frames per second is still almost perfect; many observers consider 10 frames per second acceptable for teleconferencing, but lower values (which often occur at low bit rates) may look rather like a series of still pictures than continuous motion.

In most recent codecs, the transmitted frame rate is adaptive, meaning that the algorithm automatically tries to optimize the combination of encoding accuracy and motion smoothness. However, it is possible to manually select a minimum permissible transmitted frame rate.

2.5.7 Encoding Algorithm

The above features produce a considerable but often not sufficient amount of data compression. An added feature is motion compensation, which uses frame-to-frame comparison of blocks that are not only in the same but also in adjacent locations. This way it can be determined in which direction a part of the picture has moved between frames. Motion compensation can significantly improve the quality of the picture. H.261 requires implementation of motion compensation in the decoder section of the codec, but it is optional in the encoder.

2.5.8 Bit Rates

The available (or affordable) transmission facility for a teleconference system generally determines the bit rate at which the codec can operate. This rate is manually selected. The higher the bit rate, the less data compression is needed and the higher the quality of the viewed picture.

The transmission bit rates are specified in terms of $P \times 64$ Kbps. Based on commonly available transmission facilities, the highest value of P is usually 24 in the United States and 30 in Europe. $P = 1$ corresponds to a single B channel in ISDN or a switched 56 channel. $P = 2$ corresponds to an ISDN *basic rate interface* (BRI) or a pair of switched 56 channels. $P = 6, 12$ correspond to fractional T1 circuits, while $P = 24$ is a full T1.

The resolution obtainable with QCIF is so limited that using a higher rate would be a waste of resources. A value of $P = 1$ is possible but can be used in practice only for a very low-quality picture and with G.728 audio. In practice, some transmission capacity must be reserved for audio and other ancillary functions. $P = 1$ is marginally usable with CIF if there is little motion in the picture. At $P = 6$, CIF provides acceptable picture quality for most video teleconferencing applications. Higher rates ($P = 12, 24$) produce pictures that, though still below broadcast quality, are judged only slightly degraded by most observers.

The state-of-the-art in video compression technology is in a steady flux, and new developments result in ever-improving performance. This allows the use of lower and lower bit rates not only for videophone but for many other video teleconferencing applications. At present there is the objective to be able to use ordinary telephone circuits for some videophone applications.

2.5.9 Audio

Audio is a vital part of any teleconference. Indeed, it is generally considered more important than video since its lack would destroy the usefulness of the conference.

Furthermore, audio teleconferences have been around for many years and participants have become accustomed to excellent quality. Therefore, an adequate portion of the available transmission bit rate must be reserved for audio.

There is no single audio transmission standard assigned for use with the H.261 video codec. Table 2.3 lists four available choices. The "G" standards are officially established by the *International Telephone and Telegraph*

Table 2.3
Audio Coding Standards

CCITT Recommendation	*BW*	*Bit Rate*	*Coding Algorithm*
G.711	3 KHz	64 Kbps	PCM
G.722	7 KHz	48, 56, 64 Kbps	Dual band, DPCM
G.728	3 KHz	16 Kbps	Low delay code excited linear prediction
AV.253	7 KHz	32 Kbps	

Consultative Committee (CCITT). With a low available total bit rate it is obviously important to keep the requirement for audio as low as possible. At 384 Kbps or above, it is probably feasible to assign 48 Kbps or 64 Kbps to audio, resulting in high quality without imposing an undue constraint on video.

2.5.10 Bandwidth

The wider the audio bandwidth, the more intelligible and natural speech becomes. The 3-KHz bandwidth shown in Table 2.3 for G.711 and G.728 corresponds to toll quality, similar to that experienced in normal analog telephone service, and provides good intelligibility. The 7-KHz bandwidth for G.722 and AV.253 is higher quality and will provide more natural sounding speech but is not as good as FM radio or CD quality. Most users of VTC will definitely prefer the higher quality audio.

2.5.11 Lip Sync

An important feature of the digital audio system is its inherent transmission delay. The different processing to which video and audio signals are subjected basically produces different delays. For a video teleconference close lip sync is desirable. Any noticeable deviation can be extremely annoying to all participants. Therefore, the processing delays of the video and audio channels should be accurately equalized.

2.5.12 Echo Cancellation

Many rooms contain multiple microphones distributed around a conference table. The audio from these microphones can be combined in three different ways to form one signal for transmission. They can be mixed, manually switched, or automatically switched according to voice activity.

An objectionable echo can occur when voice signals are sent back to the originating end as a result of speaker-to-microphone coupling. The echo problem is solved by two different approaches, namely, suppression or cancellation. Echo suppressers automatically reduce the level of the return audio path when a local participant is talking, thereby reducing or eliminating the associated echo. Echo cancelers generate a pseudo-echo based upon the incoming audio signal and subtract this pseudo-echo from the transmitted signal containing the echo. In general, echo cancelers are more complex than echo suppressers, and they typically provide higher voice quality. Echo suppressers can exhibit a voice clipping characteristic that in some cases can be disturbing.

The codec used for teleconferencing provides for continuous audio transmission in both directions. This is called full-duplex. However, with a typical room configuration the sound from the far end of the room is radiated from the room loudspeakers and finds its way, either directly or by means of reflections, to the microphones. This causes an echo to be heard in the far end of the room of the sounds originating there. This is a very annoying effect and should not be tolerated. There are three basic solutions to this problem.

The simplest is to use a conventional telephone handset. The configuration of the handset effectively decouples the loudspeaker and microphone. However, this approach is feasible only for single-user videophones and even there is not desirable.

Another approach is to sense in which room speaking is actually taking place at any instant of time and only allow sounds to be transmitted from that room. Since this approach allows transmission in only one direction at a time, it is known as half-duplex, or echo gating. This technique is widely used in speakerphones. However, it is not completely satisfactory for VTC because of the increased delay involved. Conversations are awkward and whole words or phrases may be completely lost due to switching.

By far the best approach is to use echo cancellation. In this technique the incoming and outgoing audio signals are compared electronically, and the portion of the outgoing signal that is due to the incoming signal is subtracted from it, ideally leaving only that portion of the signal that actually originated in the near end of the room. While cancellation is not perfect, most echo cancelers do a very good job.

2.5.13 Error Control

Digital transmission is relatively tolerant to transmission errors, but they still are an important cause of picture defects. There is usually a rather sharp error threshold between good performance and failure. With interframe coding, the effect of uncorrected errors can persist for quite some time.

H.261 requires the addition of the FEC bits in the encoder but does not mandate their use in the decoder. Since the bits have already been used in the transmission, it makes sense to use them in the decoder. How much FEC will improve performance depends on the bit rate, the error rate, and the type of errors produced by the network. For random bit error rates of about 10^{-8} or better, the picture will appear virtually errorfree even without FEC. Between 10^{-8} BER and 10^{-3} BER the FEC will improve performance but will be quite poor at 10^{-4}. At error rates poorer than 10^{-3} the system will probably be unusable.

2.5.14 Performance Impairments

The various techniques available to achieve a large amount of data compression inevitably result in an impaired picture. The user has to determine what amount and type of impairment is tolerable for his particular application of the teleconferencing system. This will provide guidance for the selection of terminal equipment and transmission facilities. The following is a listing and brief description of the major impairments.

- *Blurriness:* There are two types of blurriness. One is static and is mainly caused by limited resolution due to the picture sampling process, particularly with QCIF. Limitations imposed by the intraframe encoding procedure also add to this impairment. The other type is best described as smearing of moving objects. It is caused by limitations in the interframe coding, which makes it impossible for the encoded picture to follow changes with sufficient speed.
- *Blocking:* This impairment manifests itself as a spurious checkerboard pattern superimposed on the picture. It is caused by imperfections in the encoding of the 8- by 8-pixel block structure specified by the H.261 algorithm.
- *Image Retention:* Similar to the smearing of moving objects, interframe coding often does not allow fast enough updating of a sudden picture change. This shows up, for instance, as retention of a pattern after it has been erased or a noticeable delay in updating of the image after a switch between radically different pictures.
- *Jerkiness:* Any motion in the picture lacks smoothness but appears as a sequence of jumps. This is caused by frame repetition, which results in a lowered average transmitted frame rate.
- *Quantizing Noise:* The quantizing process results in discrete steps that do not accurately represent the input. The difference produces a spurious signal component that appears similar to conventional noise. Generally, this impairment is only slightly noticeable.

Other types of picture impairments are often mentioned, but they are generally used in an analysis by experts. In both appearance and cause they are fairly similar to one or more of the aforementioned impairments.

2.6 PLANNING A VTC SYSTEM

This section provides the reader with some guidance on the techniques and procedures to plan for the introduction of a VTC system. We discuss steps in VTC analysis such as user survey, schedule, and communication.

2.6.1 User Survey

The first step in planning a VTC room or multisite system is to survey potential users of the VTC system. The primary purpose of the survey is to determine the amount of travel used by personnel in the organization. In general, organizations whose personnel travel extensively will benefit most from VTC. The survey would be implemented using a questionnaire that would obtain:

- Number of trips per year;
- Number of persons per trip;
- Trip duration;
- Level of participants;
- Destinations.

2.6.2 Develop a Technical Approach

The second step in the planning process is the development of the technical approach for the implementation of the VTC system. Decisions that must be made are:

- Room size, which depends largely on the typical number of participants per conference;
- Electronic equipment, such as the number and size of TV monitors, cameras, graphics scanner, audio equipment, control system, and possible encryption;
- Transmission bit rate, which determines picture and audio quality;
- Custom room, rollabout, or desktop configuration;
- Multipoint requirements—if a multisite system is being developed.

2.6.3 Cost Estimate

The next step is to estimate the cost of the VTC system. The elements of the system for which cost data must be developed are nonrecurring, such as:

- Electronic equipment;
- One-time communication access;
- Furniture;
- Room modification, if any;
- Installation.

and recurring, such as:

- Communication;
- Maintenance;
- Administration, training, and operation.

2.6.4 Schedule

An important part of the planning process is the establishment of a schedule for the VTC project. Important elements of the schedule are:

- Survey;
- Development of a technical approach;
- Cost estimate;
- Cost/benefit analysis;
- Finalization of the plan and proposal;
- Decision to (not to) implement;
- Ordering materials;
- Ordering communications;
- Employment and training of staff;
- Installation;
- System test;
- User indoctrination;
- Project complete.

2.6.5 Project Proposal

Assuming the results of the cost/benefit analysis are positive, the final step in the planning process is to organize all the planning data into a document that describes the proposed project. An example of an outline of a proposal is as follows.

1. Executive summary;
2. Requirement analysis and survey;
3. Implementation plan;
4. Cost estimate;
5. Cost/benefit analysis;
6. Schedule;
7. Conclusions.

Video Compression Techniques 3

3.1 OVERVIEW

Figure 3.1 is a functional block diagram of a generic system that digitally transmits video over a communication channel. At the transmitter, the input analog signal is first filtered such that the upper cut-off frequency of the signal is N cycles/s. The filtered signal is next sampled at a rate of at least $2N$ samples per second (the Nyquist rate) to avoid aliasing distortion. Each sample is defined as a pixel (picture element) that is commonly encoded with 8-bit accuracy because this precision is required to avoid any visible distortion in the output image. At this point the bit rate is typically $16N$ bits/s, which may exceed the bit rate of the transmission channel (C bits/s). The purpose of the compressor is to reduce the $16N$ bit rate by reducing the pixel-to-pixel redundancy inherent in the image. The channel coder (for example, modem) processes the binary compressed signal for efficient transmission over the communication channel. The compressor is commonly referred to as a *source* coder (signal source) as contrasted with the *channel* coding process. As shown in Figure 3.1, the functions at the receiver are the inverses of those at the transmitter.

The ITU H.261 video coding standard is universally used for virtually all video teleconferencing systems. The picture format most commonly used for videoconferencing is the Full CIF, which is defined in Table 3.1.

If the picture rate is 30 frames/s and each input pixel is encoded to 8-bit precision, the resultant bit rate of the uncompressed signal to be coded for transmission is 36.5 Mbps. Transmission bit rates for typical communication channels are 1.544 Mbps (a T1 channel), 384 Kbps, and 64 Kbps. The purpose of the video coder is to compress the input signal so that it can be transmitted through channels of this type. The required compression ratios for these transmission data rates are 24.3, 95, and 570 to 1, respectively. The compression problem is clearly a formidable one.

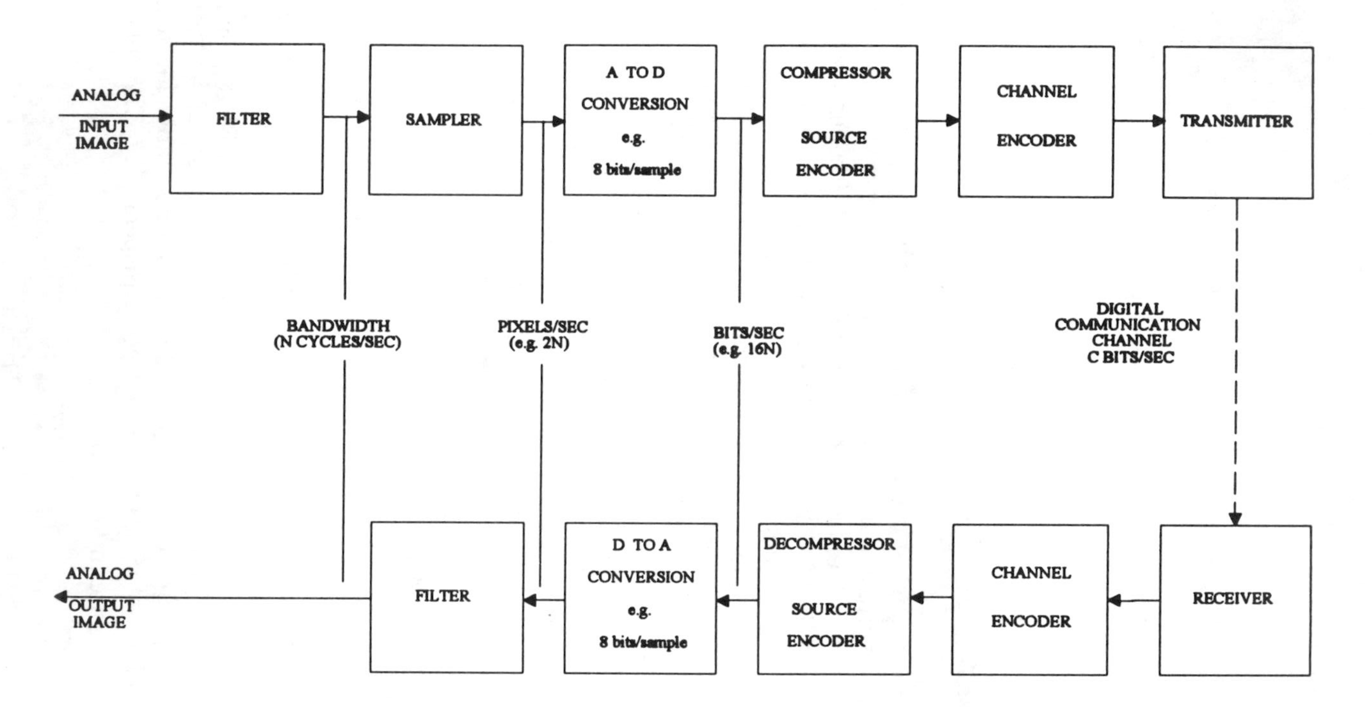

Figure 3.1 A generic system for the digital transmission of images.

Table 3.1
Pixels/Frame for CIF Picture

	Pixels/Line	*Lines/Picture*	*Total Pixels*
Luminance	352	288	101,376
Chroma R-Y	176	144	25,344
Chroma B-Y	176	144	25,344
			152,064

3.2 A GENERIC VIEW OF VIDEO COMPRESSION

Figure 3.2 is a functional block diagram of a generic video compression system. It shows that the compressor is typically implemented in three sequential operations, namely, signal analysis, quantization, and variable length coding. Basically, the signal analyzer performs measurements on the input uncompressed pixels. For example, the analyzer may compute prediction errors, compute transform coefficients, filter the signal into sub-bands, correlate the pixels with prestored VQ patterns, or correlate the pixels with the image itself (fractals). The analyzer performs measurements within a single frame (intraframe) and/or from frame to frame (interframe). Typically, no compression is achieved by the signal analysis function. The input pixel data is merely transformed into another format that is more compressible than the original signal format. For example, in the DCT case, a block of 8 × 8 pixels is converted into an 8 × 8 array of DCT coefficients.

The output of the typical signal analyzer process is usually quite accurate—with 8-bit to 12-bit precision. In lossless compressers, this accuracy is preserved in the quantization process. In lossy systems, the quantizer reduces the accuracy of the transformer output in a way that is as acceptable to the eye as possible. The quantization can be employed on either a scalar or vector basis. Most of the compression is achieved in this step by coarsely quantizing the transformed signal in such a way that any distortion to the eye is minimized. Typically, large numbers of coefficients are discarded because their value is low.

The final step in the compression process is to encode the quantizer output with a *variable length code* (VLC, sometimes called entropy coding). VLC is a technique whereby each event is assigned a code that may have a different number of bits. To obtain compression, short codes are assigned to frequently occurring events and long codes are assigned to infrequent events. The expectation is that the *average* code length will be less than the fixed code that would otherwise be required. A major advantage of VLC is that it does not degrade

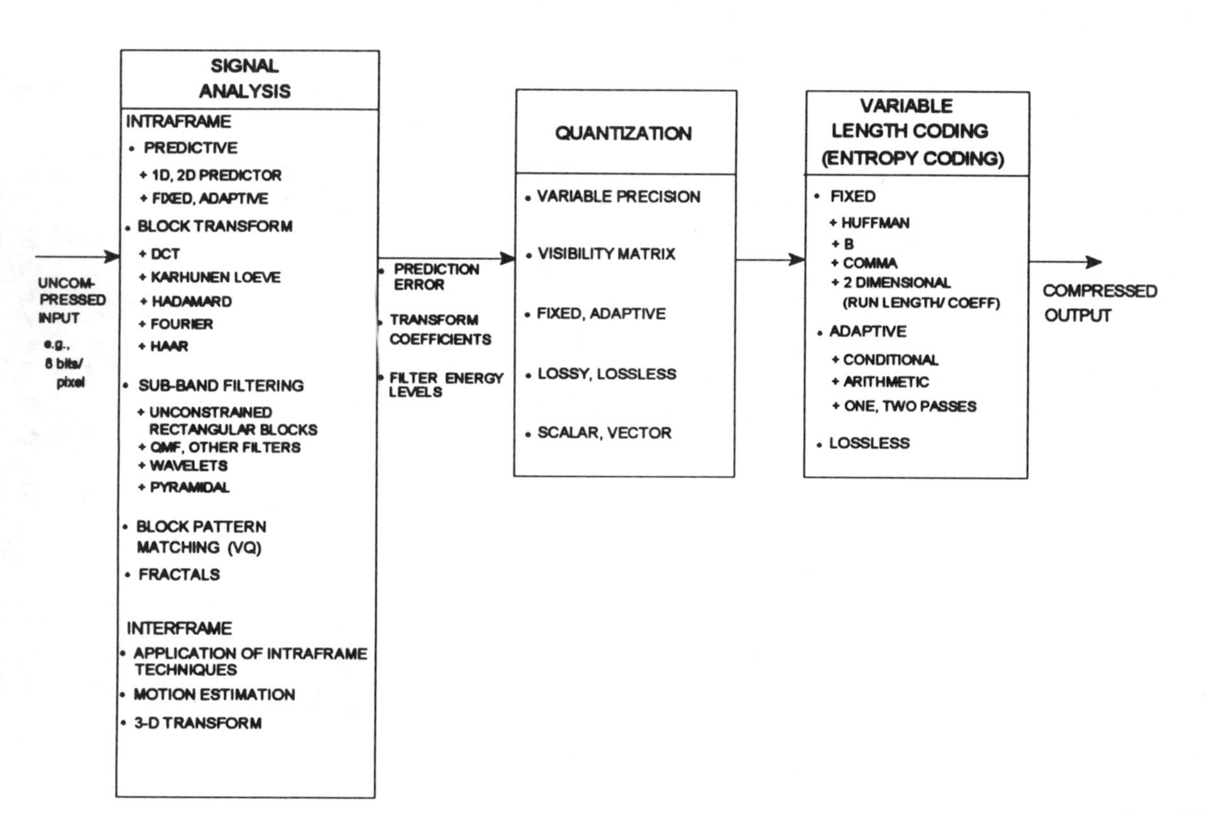

Figure 3.2 Functional block diagram of a video compression system.

the signal quality in any way (lossless). Therefore, the output picture quality is transparent to the VLC used.

3.2.1 Coding Techniques

The efficient coding of video signals has been the subject of research and development for many years, and the number of different compression algorithms that has been studied is extremely large. The most well-known coding techniques of the past, present, and future are listed in Table 3.2. The first two—namely, *pulse code modulation* (PCM) and *predictive* (differential PCM)—are very basic coding techniques. PCM provides no compression and is used as the reference to measure the performance of coding techniques. DPCM is still used today for the lossless coding of still pictures (for example, JPEG). Transform coding is the most dominant technique used to transmit video today. The *discrete cosine transform* (DCT) is used in the H.261 standard, MPEG, HDTV, and H.263 standards. The next three techniques—namely, vector quantization, wavelets, fractals, and object-based coding—can be considered *waveform* coding techniques that are competitive with the DCT. Finally, object-based coding is more advanced than the others since it isolates and encodes *objects* in the scene. Table 3.2 summarizes three key characteristics of these techniques, and each of these coding techniques is discussed in turn.

3.2.2 Pulse Code Modulation (PCM)

In PCM coding, the signal is sampled at the Nyquist rate, each sample is quantized to 2^m levels, and each level is represented by a binary word m bits in

Table 3.2
Coding Technique Overview

Coding Techniques	*Geometric Element of the Picture Being Encoded*	*Content Based*	*Symmetry**
PCM	Pixel	No	S
Predictive (differential PCM)	Pixel	No	S
Transform	Square Block of Pixels	No	M
Vector quantization	Square Block of Pixels	No	MM
Wavelet	Multiresolution Filtered Elements	No	M
Fractal	Block having any size or shape	Yes	MM
Object-based coding	Moving Objects	Yes	MM

*Complexity of the Encoder relative to the Decoder: S, essentially the same; M, more complex; MM, much more complex.

length. At the decoder, these binary words are converted to a series of amplitude levels that is low-pass filtered. Since each sample is encoded independently of its neighbor, no redundancy is reduced and PCM is used as reference to measure the performance of bit-rate reduction techniques. The PCM coding precision that is typically used for this reference is 8 bits per pixel because this precision assures that there is no visible degradation in the picture. Figures 3.3, 3.4, and 3.5 illustrate the types of contouring distortion that occur when the coding precision is limited to 2-bit, 3-bit, and 4-bit precision, respectively.

The eye is logarithmic in its sensitivity to brightness distortion; that is, it is equally sensitive to equal percentage changes in brightness. For this reason the PCM quantization is typically nonlinear with the size of the quantizer steps increasing with brightness. This nonlinear companding improves the picture quality for a given number of bits per pixel.

3.2.3 Predictive Coding

PCM transmits each pel as an independent sample without taking advantage of the high degree of pixel-to-pixel correlation existing in most pictures. Predictive coding is a basic bit-rate reduction technique that does reduce this pel-to-pel

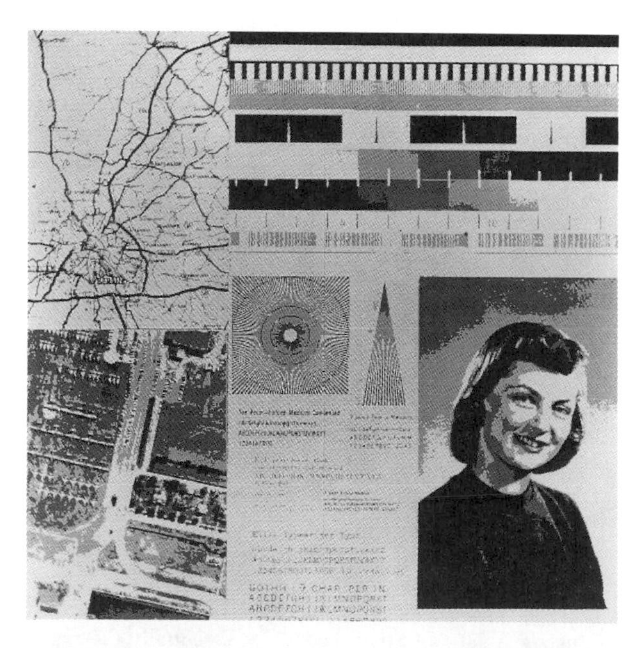

Figure 3.3 Two-bit PCM encoding.

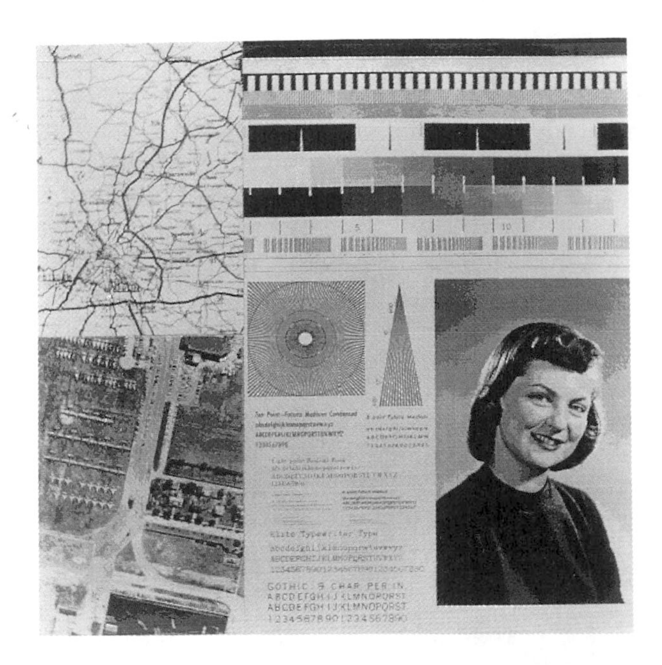

Figure 3.4 Three-bit PCM encoding.

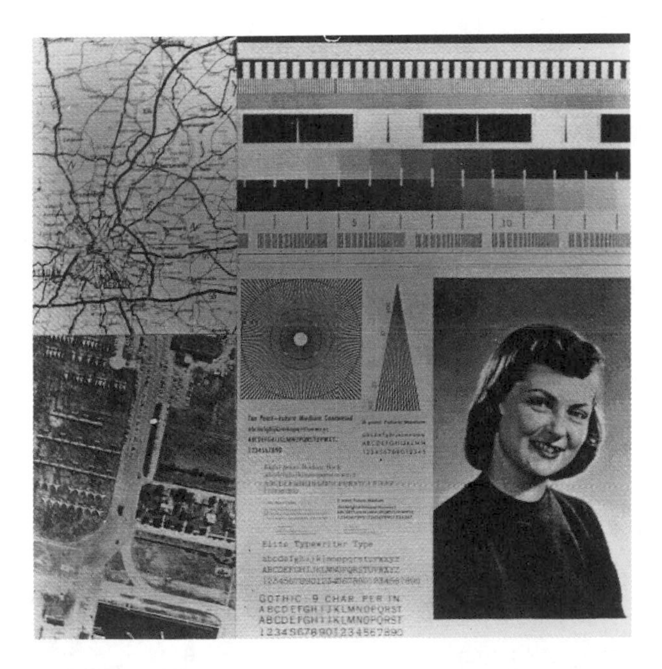

Figure 3.5 Four-bit PCM encoding.

redundancy. Figure 3.6 is a block diagram illustrating the basic predictive coding process. A predictor predicts the brightness value of each new pel based solely upon the pels that have been previously quantized and transmitted. The predicted brightness value is subtracted from the actual brightness value of the new pel resulting in a bipolar prediction error signal. This error signal is quantized and transmitted. This quantization process can vary over a wide range of complexities. The most common technique is conventional pel-by-pel scalar quantization. However, it is also possible to employ transform coding or vector quantization. The quantization can be fixed, or it can adapt to the data itself. The quantizer can also vary over a wide range of accuracies. If one bit quantization is employed, the system becomes the well-known *delta modulation* technique. If the predictive quantizer employs multiple bits per pel, the technique is commonly defined as *differential PCM* (DPCM). At the receiver the inverse of the quantization process is performed and the decoded error signal is added to the predicted value to form the output signal for viewing. The output signal is fed to the predictor to be used for prediction of the next pel. Referring back to the predictive encoder the reader will note that the transmitted signal is decoded

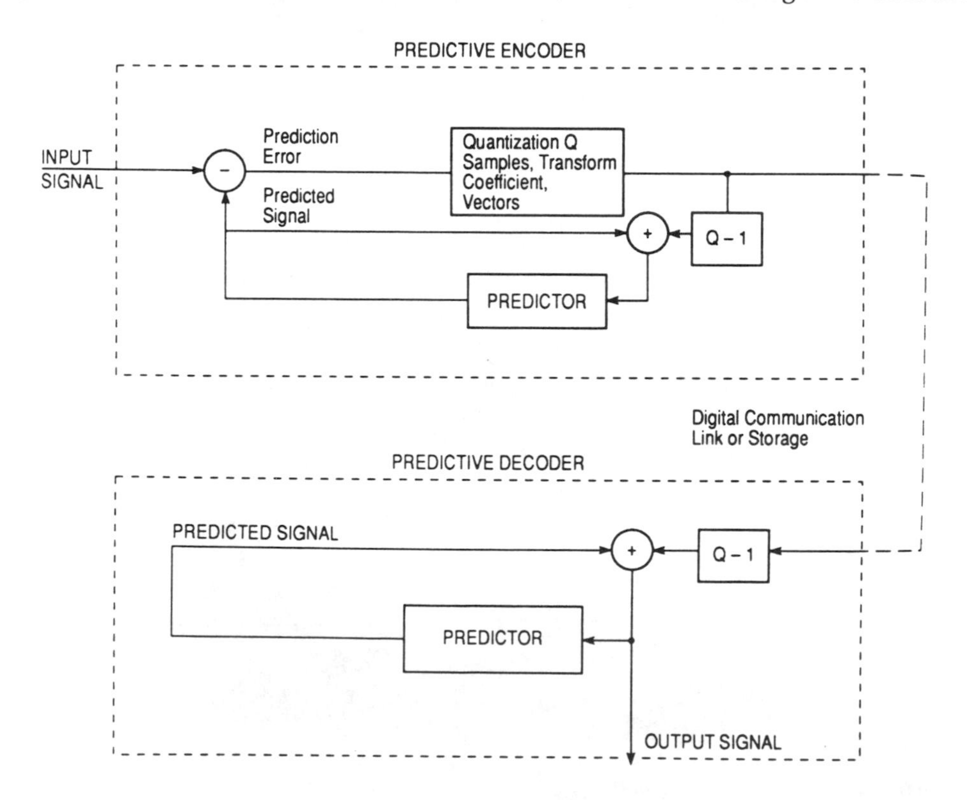

Figure 3.6 Block diagram of a generic predictive coding system.

at the transmitter using exactly the same decoding process that is used at the receiver. The predictive encoder can be viewed as a servo loop that continually forces the decoded output signal to be as close as possible to the input signal.

Figure 3.7 illustrates the transfer function of a typical 3-bit DPCM predictive coder. The quantizer is usually nonlinear because the eye is very sensitive to small changes in low-detail portions of a picture (small prediction error),but the eye is insensitive to coarse quantization of high contrast edges (large predictive error). The design of this quantizer is a compromise between conflicting objectives. It is desirable that the quantizer precision be fine, particularly for small error signals, to keep the background granular noise in the output picture at an acceptably low level. On the other hand, the quantizer steps must be large enough, particularly the largest increment, so that the output can respond reasonably well to high-contrast changes in the input picture. If the largest increment is too small, slope overload occurs, resulting in picture blurring.

3.2.4 Transform Coding

Transform coding algorithms, generally speaking, operate as two-step processes. In the first step a linear transformation of the original signal (separated into

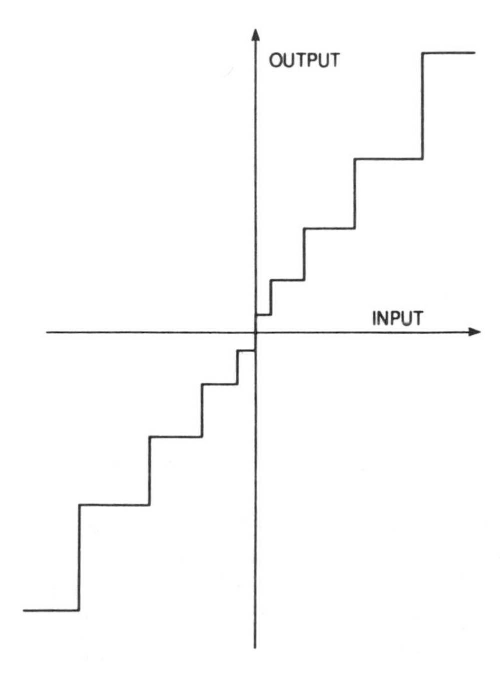

Figure 3.7 Quantizer for predictive encoder.

sub-blocks of $N \times N$ pels each) is performed, in which signal space is mapped into transform space. In the second step, the transformed signal is compressed by encoding each sub-block through quantization. The reconstruction operation involves performing an inverse transformation of each decoded transformed sub-block. The function of the transformation operation is to make the transformed samples more independent than the original samples so that the subsequent operation of quantization may be done more efficiently.

The transformation operation itself does not provide compression; rather, it is a remapping of the signal into another domain in which compression can be achieved more effectively. Compression can be achieved for two reasons. First, not all of the transform domain coefficients need to be transmitted to achieve acceptable picture quantity. Second, the coefficients that are transmitted can be encoded with reduced precision without seriously affecting image quality.

3.2.5 Transformation Techniques

Transforms that have proven useful include the Karhunen-Loeve, discrete Fourier, discrete cosine, and Walsh-Hadamard transforms. The *Karhunen-Loeve transform* (KLT) is considered to be an optimum transformation, and for this reason many other transformations have been compared to it in terms of performance. However, the KLT has certain characteristics that make it less than ideal for image processing. For data having high interelement correlation, the performance of other transforms (such as the discrete cosine transform) is virtually indistinguishable from that of the KLT and, thus, usually does not warrant its added complexity.

The discrete Fourier transform is one of the few complex transforms used in data coding schemes. There are disadvantages to using a complex transform for data coding, the most obvious of which is the storage and manipulation of complex numbers. Again, as in the case of the KLT, this complexity issue would not be a factor if the performance of the DFT was significantly greater than that of other transforms. However, other transforms that are less complex perform better than the DFT.

The *discrete cosine transform* (DCT) is one of an extensive family of sinusoidal transforms. In their discrete form, the basis vectors consist of sampled values of sinusoidal or cosinusoidal functions that, unlike those of the DFT, are real number quantities. The DCT has been singled out for special attention by workers in the image processing field, principally because, for conventional image data having reasonably high interelement correlation, the DCT's performance is virtually indistinguishable from that of other transforms that are much more complex to implement.

The three transforms mentioned previously have basis functions that are either cosinusoidal (that is, the Fourier and discrete cosine) or are a good approximation of a sinusoidal function, such as the Karhunen-Loeve transform. The Walsh-Hadamard transform is an approximation of a rectangular orthonomal function. The actual transform consists of a matrix of +1 and −1 values, which eliminates multiplications from the transform process. The elimination of multiplications is a significant property, since the aforementioned transforms require real or complex multiplications. However, the Walsh-Hadamard transform does not provide the excellent performance that the discrete cosine transform provides.

Since the DCT is universally accepted as the preferred transform for image coding, it is useful to provide more detail on its implementation. The execution of the DCT algorithm requires the division of an image into a series of $(N \times N)$ sub-blocks of pixels. Each sub-block is transformed by a two-dimensional $(N \times N)$ DCT process as $[T] = [C] + [D] = [C]^T$ where $[T]$ is the transformed sub-block, $[C]$ is the DCT basis matrix, and $[D]$ is the input data sub-block ($[C]^T$ is the transpose of the DCT basis matrix). The DCT basis matrix coefficients were determined from the relation

$$C_{1 \cdot j} = C_o \cdot (2/N) \cdot (\cos(i \cdot (j + 0.5) \cdot (n/N)))$$

where $C_o = 1/2$ for $1 = 0$, $C_o = 1$ otherwise, and $i = j = 0$ to $N - 1$. Figure 3.8 illustrates the basis functions for a 8×8 DCT. This transformation converts each $(N \times N)$ sub-block of pixels into an $(N \times N)$ matrix of transform coefficients, which consists of one DC coefficient and $(N \times N - 1)$ AC coefficients. The sum of the squares of all of the AC coefficients in a given transform matrix is known as the AC energy of that transform matrix.

3.2.6 Coding of Transform Coefficients

As explained in the previous section the DCT is usually used when pictures are transmitted using transform techniques. This transformation merely creates a set of coefficients equal in number to the original set of pels. At this point no compression has been accomplished except that the original set of pels with uniformly high redundancy have been decorrelated and the information has been compacted in the lower spatial frequency coefficients. The purpose of this section is to address the second part of the two-step process, that is, how to encode the transform coefficients for transmission.

The first step in the coding process is to determine which coefficients are to be transmitted and which are to be deleted. Figure 3.8 illustrates the set of 64 transform coefficients corresponding to a 8×8 block of pels to be coded. Coefficient number one is the DC coefficient that is a measure of the average

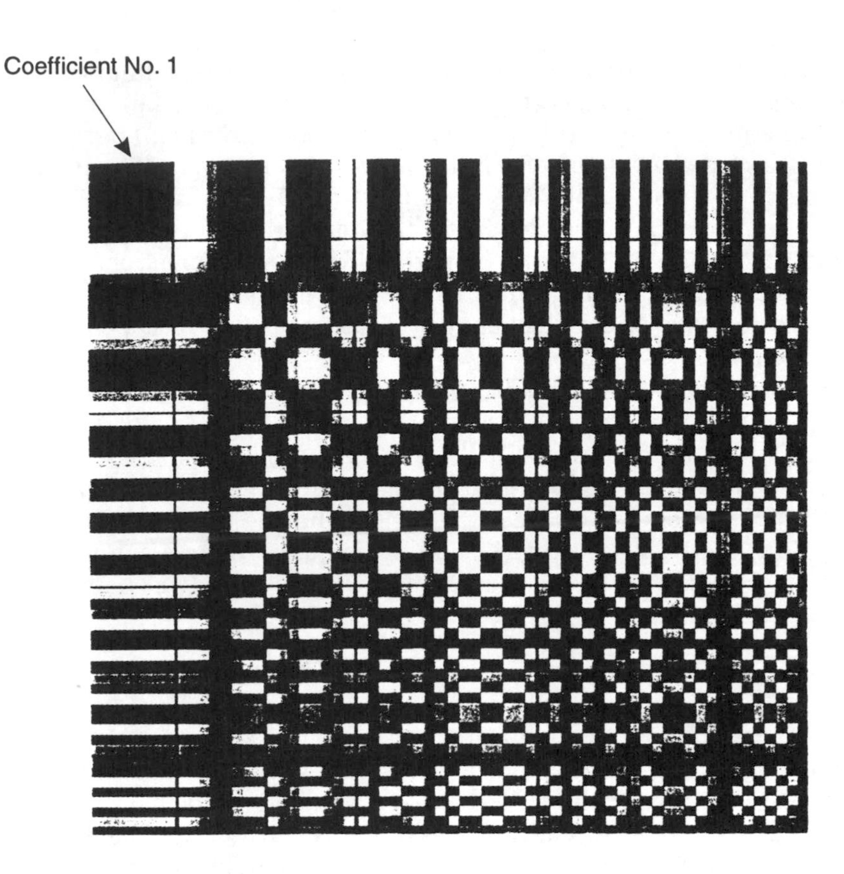

Figure 3.8 DCT basis function.

brightness of the block. Coefficients in the top row measure of spatial frequency content in the horizontal direction. Coefficients in the left column measure frequencies in the vertical direction, and all others measure various combinations thereof. In general, most of the energy is contained in the low-frequency coefficients with relatively little signal strength in the high-frequency coefficients.

Figure 3.9 shows a simple example of how each 8 × 8 block is coded. Figure 3.9(a) shows the original block to be coded. The block has a constant slope or shading from the upper left-hand corner to the lower right. Without compression, this would take 8 bits to code each of the 64 pixels or a total of 512 bits. First, the block is transformed, using the two-dimensional DCT, giving

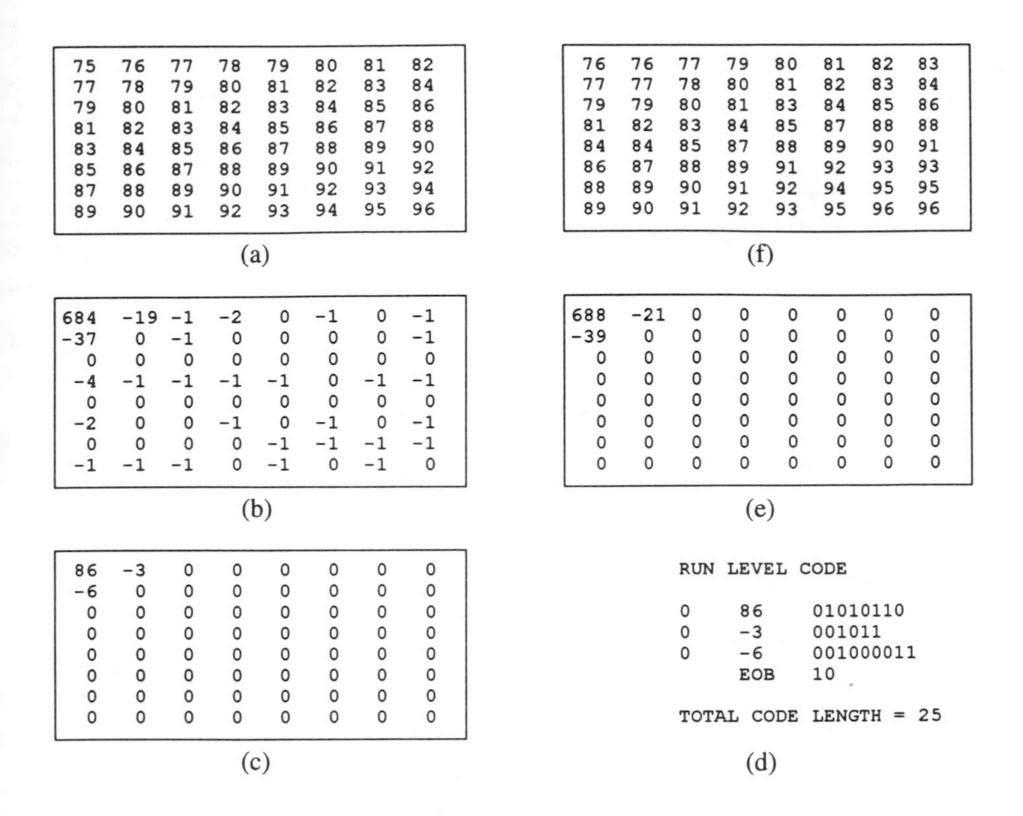

75	76	77	78	79	80	81	82
77	78	79	80	81	82	83	84
79	80	81	82	83	84	85	86
81	82	83	84	85	86	87	88
83	84	85	86	87	88	89	90
85	86	87	88	89	90	91	92
87	88	89	90	91	92	93	94
89	90	91	92	93	94	95	96

(a)

76	76	77	79	80	81	82	83
77	77	78	80	81	82	83	84
79	79	80	81	83	84	85	86
81	82	83	84	85	87	88	88
84	84	85	87	88	89	90	91
86	87	88	89	91	92	93	93
88	89	90	91	92	94	95	95
89	90	91	92	93	95	96	96

(f)

684	-19	-1	-2	0	-1	0	-1
-37	0	-1	0	0	0	0	-1
0	0	0	0	0	0	0	0
-4	-1	-1	-1	-1	0	-1	-1
0	0	0	0	0	0	0	0
-2	0	0	-1	0	-1	0	-1
0	0	0	0	-1	-1	-1	-1
-1	-1	-1	0	-1	0	-1	0

(b)

688	-21	0	0	0	0	0	0
-39	0	0	0	0	0	0	0
0	0	0	0	0	0	0	0
0	0	0	0	0	0	0	0
0	0	0	0	0	0	0	0
0	0	0	0	0	0	0	0
0	0	0	0	0	0	0	0
0	0	0	0	0	0	0	0

(e)

86	-3	0	0	0	0	0	0
-6	0	0	0	0	0	0	0
0	0	0	0	0	0	0	0
0	0	0	0	0	0	0	0
0	0	0	0	0	0	0	0
0	0	0	0	0	0	0	0
0	0	0	0	0	0	0	0
0	0	0	0	0	0	0	0

(c)

RUN	LEVEL	CODE
0	86	01010110
0	-3	001011
0	-6	001000011
	EOB	10

TOTAL CODE LENGTH = 25

(d)

Figure 3.9 Sample block coding: (a) original block ($8 \times 8 \times 8 = 512$ bits); (b) transformed block coefficients; (c) quantized coefficient levels; (d) coefficients in zig-zag order and variable length code; (e) inverse quantized coefficients; and (f) reconstituted block.

the coefficients of Figure 3.9(b). Note that most of the energy is concentrated into the upper left-hand corner of the coefficient matrix.

Essentially, the DCT is performed by multiplying the input block by each of the 64 basis functions shown graphically in Figure 3.8. The results of each of these multiplications, also 8×8 arrays, are summed to give the 64 transform coefficients. In the upper left-hand corner of Figure 3.8, the first basis function is constant over the block and therefore gives rise to the DC value of the input block. At the opposite corner, the basis function is a checkerboard and will give significant coefficient values only if there are elements of this pattern in the input block. Of course, the coefficients are in practice calculated by a chip in a more efficient manner than described here.

Next, the coefficients of Figure 3.9(b) are quantized with a stepsize of 6. (The first term {DC} always uses a stepsize of 8.) This produces the values of

Figure 3.9(c), which are much smaller in magnitude than the original coefficients and most of the coefficients become zero. The larger the stepsize, the smaller the values produced, resulting in more compression.

The coefficients are then reordered, using the zig-zag scanning order of Figure 3.10. All zero coefficients are replaced with a count of the number of zeros before each nonzero coefficient (RUN). Each combination of RUN and VALUE produces a VLC that is sent to the decoder. The last nonzero VALUE is followed by an *End of Block* (EoB) code. The total number of bits used to describe the block is 25, a compression of 20:1. The code in Figure 3.9(d) is based upon a two-dimensional VLC (CCITT Recommendation H.261) shown in Table 3.3.

Additional examples in Table 3.4 illustrate the coding process. In examples 1, 2, and 3, the codes are obtained directly from Table 3.3. For examples 4, 5, 6, and 7, the combinations of run length and level cannot be found in Table 3.3, and so the escape code is used, together with the binary representation of the run length (6 bits) and the 2s complement representation of the level.

At the decoder, the stepsize and VALUEs are used to reconstruct the inverse quantized coefficients, which, as shown in Figure 3.9(e), are similar to but not exactly equal to the original coefficients. When these coefficients are inverse transformed, the result of Figure 3.9(f) is obtained. Note that the differences between this block and the original block are quite small.

Figure 3.11 shows a slightly different example that shows more clearly some of the features of DCT coding. In this example, in addition to shading, the block contains a checkerboard pattern that matches the highest order basis function. This causes the last coefficient to be transmitted. There are 60 zero-value coefficients (in the zig-zag order) between the previous nonzero coefficient

1	2	6	7	15	16	28	29
3	5	8	14	17	27	30	43
4	9	13	18	26	31	42	44
10	12	19	25	32	41	45	54
11	20	24	33	40	46	53	55
21	23	34	39	47	52	56	61
22	35	38	48	51	57	60	62
36	37	49	50	58	59	63	64

Figure 3.10 Scanning order in a block.

Table 3.3
Two-Dimensional VLC Table

Run	*Level*	*Code*
EOB	1	10
0	1	1s If first coefficient in block
0	1	11s Not first coefficient in block
0	2	0100 s
0	3	0010 ls
0	4	000 110s
0	5	0010 0110 s
0	6	0010 0001 s
0	8	0000 0001 1101 s
0	9	0000 0001 1000 s
0	10	0000 0001 0011 s
0	11	0000 0001 0000 s
0	12	0000 0000 1101 0s
0	13	0000 0000 1100 1s
0	14	0000 0000 1100 0s
0	15	0000 0000 1011 1s
1	1	011s
1	2	0001 10s
1	3	0010 0101 s
1	4	0000 0011 00s
1	5	0000 0001 1011 s
1	6	0000 0000 1011 0s
1	7	0000 0000 1010 1s
2	1	0101 s
2	2	0000 100s
2	3	0000 0010 11s
2	4	0000 0001 0100 s
2	5	0000 0000 1010 0s
3	1	0011 1s
3	2	0010 0100 s
3	3	0000 0001 1100 s
3	4	0000 0000 1001 1s
4	1	0011 0s
4	2	0000 0011 11s
4	3	0000 0001 0010 s
5	1	0001 11s
5	2	0000 0010 01s
5	3	0000 0000 1001 0s
6	1	0001 01s
6	2	0000 0001 1110 s
7	1	0001 00s
7	2	0000 0001 0101 s
8	1	0000 111s
8	2	0000 0001 0001 s

Table 3.3
(Continued)

Run	*Level*	*Code*
9	1	0000 101s
9	2	0000 0000 1000 1s
10	1	0010 0111 s
10	2	0000 0000 1000 0s
11	1	0010 0011 s
12	1	0010 0010 s
13	1	0010 0000 s
14	1	0000 0011 10s
15	1	000 0011 01s
16	1	0000 0010 00s
17	1	0000 0001 1111 s
18	1	0000 0001 1010 s
19	1	0000 0001 1001 s
20	1	0000 0001 0111 s
21	1	0000 0001 0110 s
22	1	0000 0000 1111 1s
23	1	0000 0000 1111 0s
24	1	0000 0000 1110 1s
25	1	0000 0000 1110 0s
26	1	0000 0000 1101 1s
Escape	1	0000 01

Table 3.4
VLC Coding Examples

Example	*Run Length*	*Level*	*Code*
1	0	3	001010
2	0	−3	001011
3	26	−1	0000 0000 1101 11
4	32	−1	000001,100000,11111111
5	60	2	000001,111100,00000010
6	60	−2	000001,111100,11111110
7	60	−86	000001,111100,10101010

and the last one, so the run length is 60. The last coefficient is coded as a 6-bit escape code (000001), a 6-bit run code (111100) [60_{10}], and an 8-bit level code (00000010) [2_{10}] as shown at the bottom of Table 3.3.

Two types of distortion appear in transform coded pictures: truncation error and quantization errors. Quantization errors are noiselike, whereas trunca-

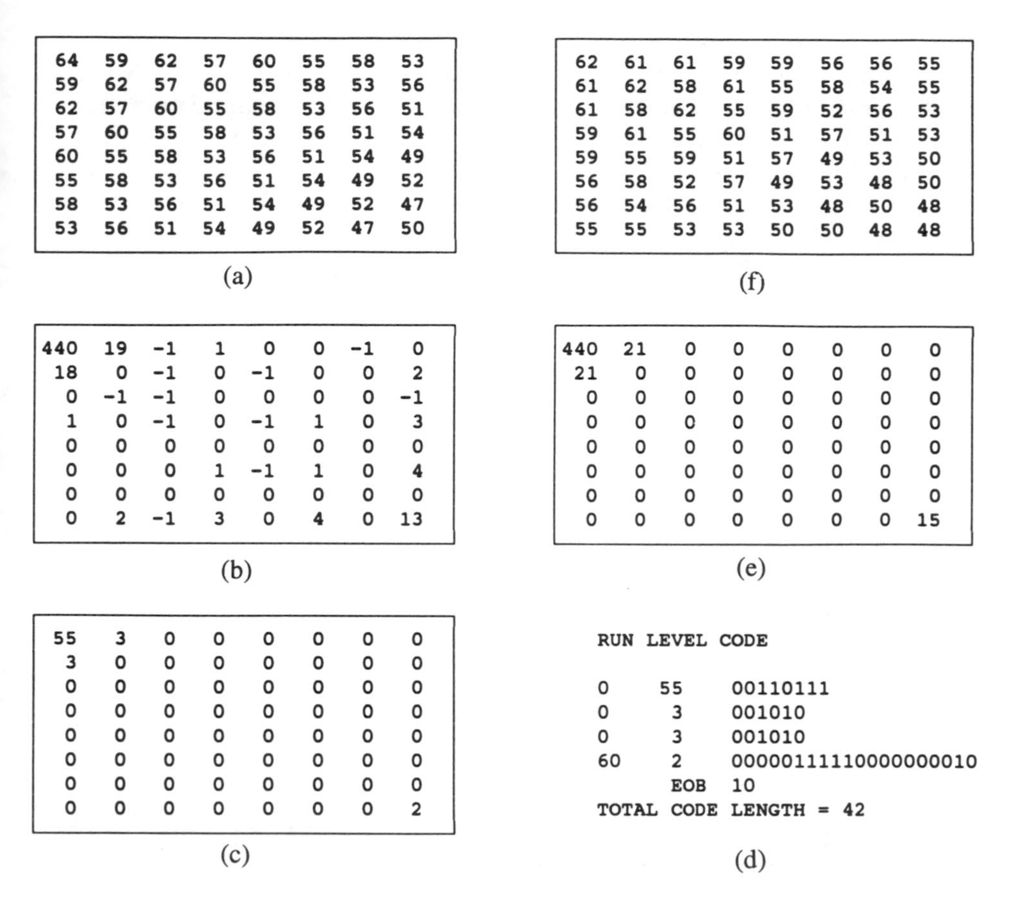

64	59	62	57	60	55	58	53
59	62	57	60	55	58	53	56
62	57	60	55	58	53	56	51
57	60	55	58	53	56	51	54
60	55	58	53	56	51	54	49
55	58	53	56	51	54	49	52
58	53	56	51	54	49	52	47
53	56	51	54	49	52	47	50

(a)

62	61	61	59	59	56	56	55
61	62	58	61	55	58	54	55
61	58	62	55	59	52	56	53
59	61	55	60	51	57	51	53
59	55	59	51	57	49	53	50
56	58	52	57	49	53	48	50
56	54	56	51	53	48	50	48
55	55	53	53	50	50	48	48

(f)

440	19	-1	1	0	0	-1	0
18	0	-1	0	-1	0	0	2
0	-1	-1	0	0	0	0	-1
1	0	-1	0	-1	1	0	3
0	0	0	0	0	0	0	0
0	0	0	1	-1	1	0	4
0	0	0	0	0	0	0	0
0	2	-1	3	0	4	0	13

(b)

440	21	0	0	0	0	0	0
21	0	0	0	0	0	0	0
0	0	0	0	0	0	0	0
0	0	0	0	0	0	0	0
0	0	0	0	0	0	0	0
0	0	0	0	0	0	0	0
0	0	0	0	0	0	0	0
0	0	0	0	0	0	0	15

(e)

55	3	0	0	0	0	0	0
3	0	0	0	0	0	0	0
0	0	0	0	0	0	0	0
0	0	0	0	0	0	0	0
0	0	0	0	0	0	0	0
0	0	0	0	0	0	0	0
0	0	0	0	0	0	0	0
0	0	0	0	0	0	0	2

(c)

RUN	LEVEL	CODE
0	55	00110111
0	3	001010
0	3	001010
60	2	0000011111000000010
	EOB	10
TOTAL	CODE	LENGTH = 42

(d)

Figure 3.11 Example of DCT coding: (a) original block, (b) transformed block coefficients, (c) quantized coefficient levels, (d) coefficients in zig-zag order and variable length code, (e) inverse quantized coefficients, and (f) reconstituted block.

tion errors cause a loss of resolution. In practice, the truncation threshold and quantization precision must be adjusted experimentally to achieve the maximum compression and acceptable picture quality. In general, transform coding is preferable to predictive coding for compression to bit rates below 1 or 2 bits per pel for single pictures. However, in those applications where cost and complexity are important issues the choice between these two algorithms may be less clear.

3.3 VECTOR QUANTIZATION

Quantization is the process by which an analog signal, having a continuous range of possible values, is divided into a limited set of discrete steps. The most common quantization procedure is *scalar*, where one can visualize a scale placed next to the variable being measured. In the scalar quantization process, one digital word represents the quantized value of one sample of a signal (for example, the value of a single pixel in an image).

In the case of *vector quantization* (VQ), one digital word represents the quantized value of more than one sample of a signal. In the case of the VQ of an image [1–9], a single word, or vector, is typically used to represent the quantized values of, for example, an entire 4 × 4 array of pixels. Unfortunately, it is difficult to visualize how one vector represents multiple quantized values in n-dimensional space. It is easier to visualize the vector quantization of an array of two adjacent pixels as illustrated in Figure 3.12. The figure shows the case where each pixel is first quantized, by conventional scalar means, to

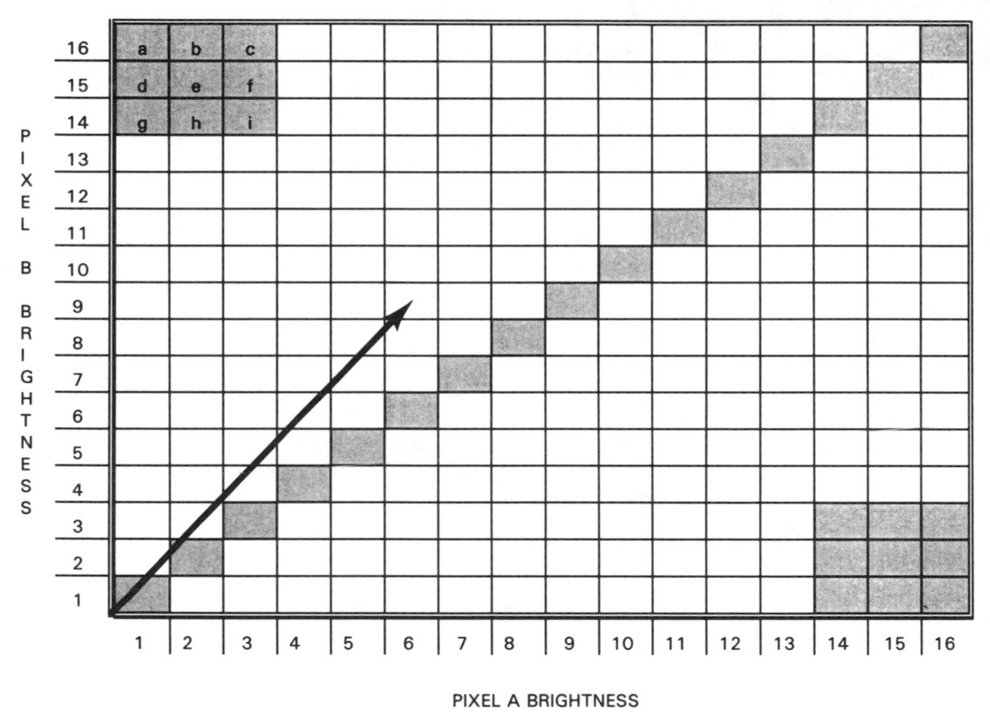

Figure 3.12 Vector quantization.

4-bit (16-level) precision. The figure illustrates one vector representing the combination of brightness value 6 for pixel A and brightness value 9 for pixel B. In general, the vector could take on $16 \times 16 = 256$ possible values; however, if only this procedure were employed, no compression would be achieved.

Compression in VQ systems is accomplished in two steps. In the first step, a number of adjacent vectors, which occur in a cluster in the input, are represented by a single vector at the center of the cluster. Clustering is accomplished using two approaches. In one case a very large sample of actual vectors from real images, usually known as a training vector set, is accumulated. From this set, vector clusters are located by rather complex algorithms, and representative vectors are placed at the centers of these clusters. The trade-off between distortion and compression is dependent on how many representative vectors are included in the list. The longer the list, the less the distortion, but also the less the compression.

In the second approach clusters are defined on the basis of psychovisual perception. For example, the eye is very insensitive to variations in high-contrast edges such as those occurring in the cluster of nine vectors (**a** through **i**) in the corner of Figure 3.12. If any of these nine vectors (**a**–**i**) occur in the input image, one vector representing the cluster of vectors could be transmitted, and brightness values corresponding to vector **e** would be displayed at the output, without any perceived distortion. Little, if any, clustering is accomplished for equal-brightness vectors (45-degree diagonal in Figure 3.12) because they occur frequently and the eye is sensitive to distortions in this region.

The lossy compression previously described merely sets the stage for the second step of compression, which is lossless VLC (also known as entropy coding). This coding takes advantage of the fact that the vectors to be transmitted are not equally likely. Short codes are assigned to the most likely vectors, and long codes are assigned to the least likely. In this way the overall VQ compression is a combination of psychovisual lossy compression and mathematical lossless compression. The above example illustrates VQ compression for a single image in two dimensions, that is, *x*, *y*. Clearly VQ can be used even more effectively to compress a video signal in three dimensions, that is, *x*, *y*, *t*.

One important characteristic of VQ is that it is highly asymmetric. That is, the encoder is far more complex than the decoder. The encoder requires a complex search process to make the decision as to what vector to transmit, while the decoder is merely a look-up table to display the value corresponding to the transmitted vector. This asymmetric characteristic makes VQ very attractive for applications where video is encoded once (for example, CD-ROM) and displayed many times on inexpensive display terminals.

3.4 WAVELET CODING

The wavelet transform [10–15] decomposes an image into frequency components. Unlike the DCT, however, the wavelet retains the spatial context of the frequency components. This decomposition is performed by iterative low- and high-pass filtering using wavelet filters. The result of the high-pass filter is retained. The result of the low-pass filter is decimated and again low- and high-pass filtered. This is continued for the number of levels desired (typically 3 or 4). See Figure 3.13.

This filtering is performed on the entire image not on small blocks like the DCT. This eliminates the blocking artifact normally associated with DCT algorithms. At each frequency band, there is a filtered version of the image whose spatial resolution is proportional to the frequency. For example, low-frequency bands have low spatial resolution that represent larger areas of the image. High-frequency bands have high spatial resolution that represent local edge and texture detail. While the DCT coefficients represent a decomposition based on frequency-related cosines, the wavelet transform is a decomposition based on frequency-related complex waveforms or wavelets. The result of the wavelet transform is a hierarchical representation of the image where each level in the hierarchy is a map of the image information contained within a specific frequency band. This hierarchy is represented by a tree structure of filtered

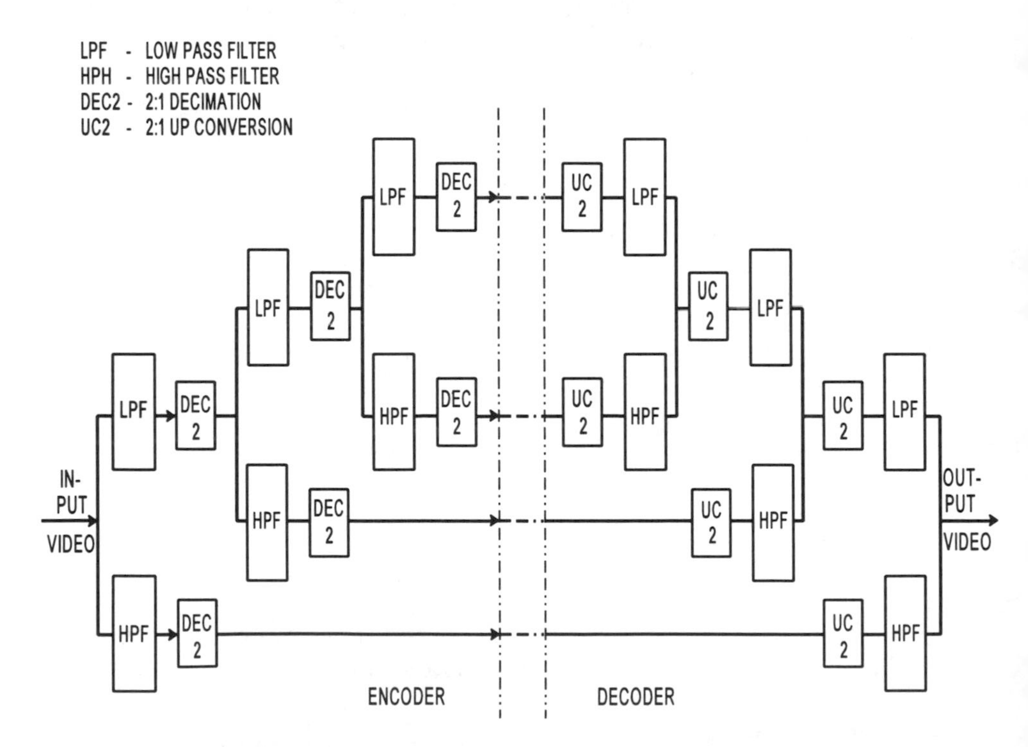

Figure 3.13 Functional block diagram of a typical wavelet codec.

pixels, with the lowest frequency pixels at the root. Each pixel at one level is linked to multiple pixels at the next level. The lowest frequency level is the called the reference signal. The other levels are the detail signals.

There are two important steps in designing a wavelet compressor: selection of the wavelet filter and effectively quantizing the resulting tree structure. The wavelet filter is divided into two parts, an analysis filter and a synthesis filter. The analysis filter is used during compression, and the synthesis filter is used during decompression. This filter pair must meet certain constraints. These are perfect reconstruction, finite length, and regularity. Perfect reconstruction indicates that a signal put into the input of the analysis filter is identical to the signal that comes out of the synthesis filter within some small error. Finite length means that it is implementable. Regularity means that the filter converges to a continuous function. Meeting these criteria, the filter is then selected to pack as much information about the original image as possible into the reference signal. This is because the reference signal is generally quantized finer than the detail signals, which are quantized coarsely or discarded.

The quantizer is the function where the compression gain is realized. The transform converts the pixels into a format that is more compressible. The quantizer does the compression. After quantizing the reference signal and detail signals, one is left with a sparsely populated tree of spatial and frequency position-dependent information. This information is further compressed by effectively representing only the significant data without losing the position information. This is done by tree trimming algorithms or run length coding. This leads to large compression ratios while maintaining image quality and detail.

In decompression, the inverse quantized, filtered reference signal, and detail signals are up sampled (interpolated) and applied to the synthesis filter banks. The outputs of these banks are summed. See Figure 3.13. The result is the original image less any distortion due to quantization.

In wavelet compression, the artifacts that occur at high compression ratios are loss of detail and ringing or oscillation around edges. The loss of detail is a result of the coarse quantization of the detail signals. The ringing is a result of the characteristics of the wavelet filters. Those filters exhibiting ringing in the impulse response of the filter will exhibit ringing to a greater extent in the resulting image.

There is work currently being done to adapt wavelet compression to motion video. There is not a single approach being used for interframe compression using wavelets. Several organizations, both commercial and academic, are investigating different techniques for interframe coding. One such technique takes the wavelet transform of the previous image and subtracts it from the wavelet transform of the current image. It then quantizes and transmits only the differences.

Another technique for motion video compression using wavelets is region-based wavelet transforms. This techniques identifies regions in the image. The regions are tracked for motion. Then wavelets are fitted to each region. Other techniques under study are adaptive wavelet transforms, overcomplete wavelet transforms, and zero crossing translates. All of these techniques attempt to account for motion components (for example, translations, rotations, and scaling) in the wavelet transform.

3.5 FRACTAL CODING

A fractal is a structure possessing similar looking forms of many different sizes. It can be magnified infinitely with structure at every scale and can be generated by small, finite sets of data and instructions. Fractal compression is based on three concepts, namely, affine maps, iterated system functions, and the collage theorem. The affine map is a combination of rotations, scalings, and translations that act on a part of a source image to create a part of a target image. An iterated function system is a collection of contractive, affine maps. The iterated function system acts on a part of the source image to create a part of the target image out of repeated parts of the source image. These repeated parts are of various rotations, scales, and translations. The collage theorem says that if an image can be described by a set of affine maps then that set of maps, provides an iterated system function that can be used to reproduce as good an approximation of the image as you desire.

The challenge is then to find a fractal model for a given image. This is done through the fractal transform, which is a systematic method that breaks up an image into smaller regions, called domain regions, and finds the best affine maps for those regions. The domain regions are nonoverlapping and completely cover the image. Range regions are also defined. They can overlap and do not need to cover the entire image. An affine map is generated that maps every range region into each domain region. The map and range region that provide the best match for a given domain region are selected. The affine coefficients, the domain region geometry, and the range region addresses for each domain range are packed to form the compressed data file, a FIF file, for still images or the compressed data stream for intra-coded motion video. See Figure 3.14.

Decompression is performed by first creating two arbitrary buffers. The domain regions are identified on one buffer and the range regions on the other. The affine maps are applied using the range buffer and filling the domain buffer. The range and domain regions are then identified on the opposite buffers and the affine transforms are applied again. This is repeated until the differences between the two buffers are sufficiently small. This process results in an image that closely matches the original image before compression. See Figure 3.15.

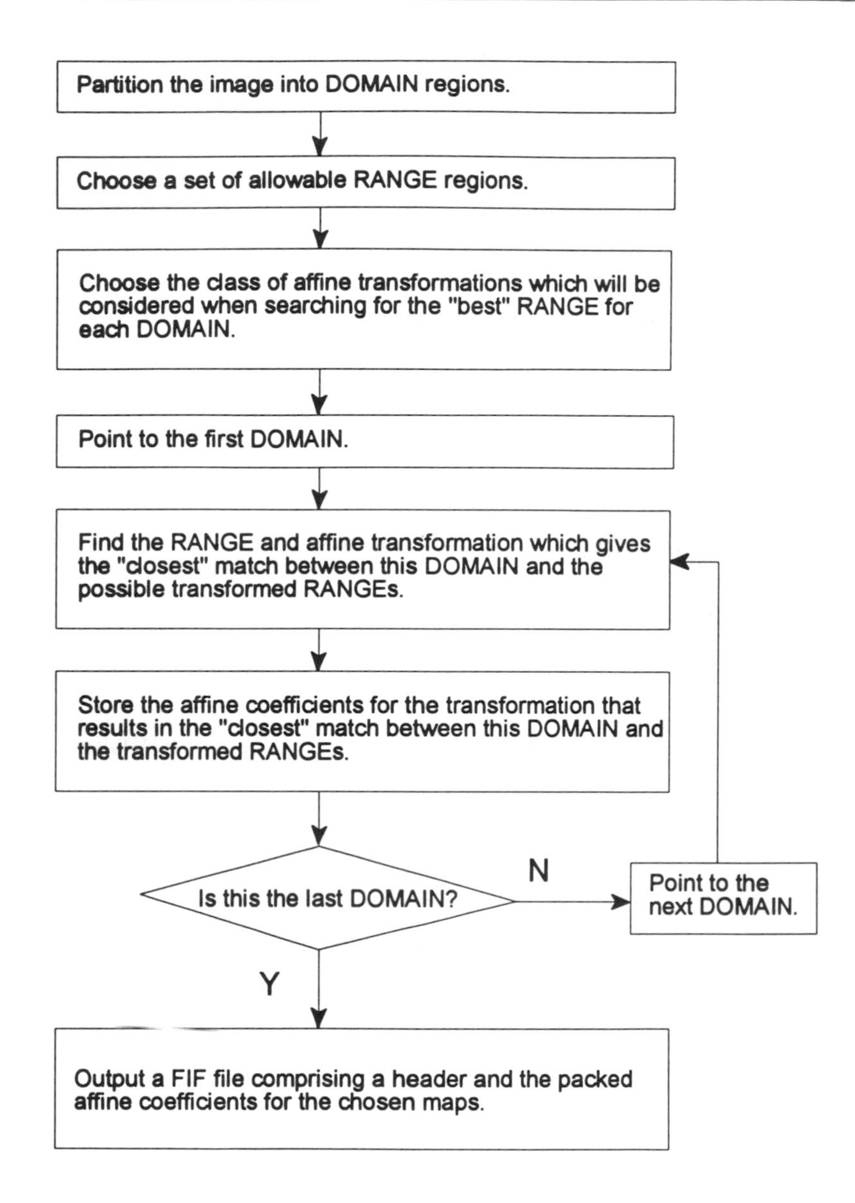

Figure 3.14 Fractal transform still-image compression.

Fractal coding of the difference images typically used in interframe coding does not work well because of the statistics of the difference image. One technique that is under study uses the difference image to identify areas where the prediction fails. The areas where prediction was satisfactory are defined as

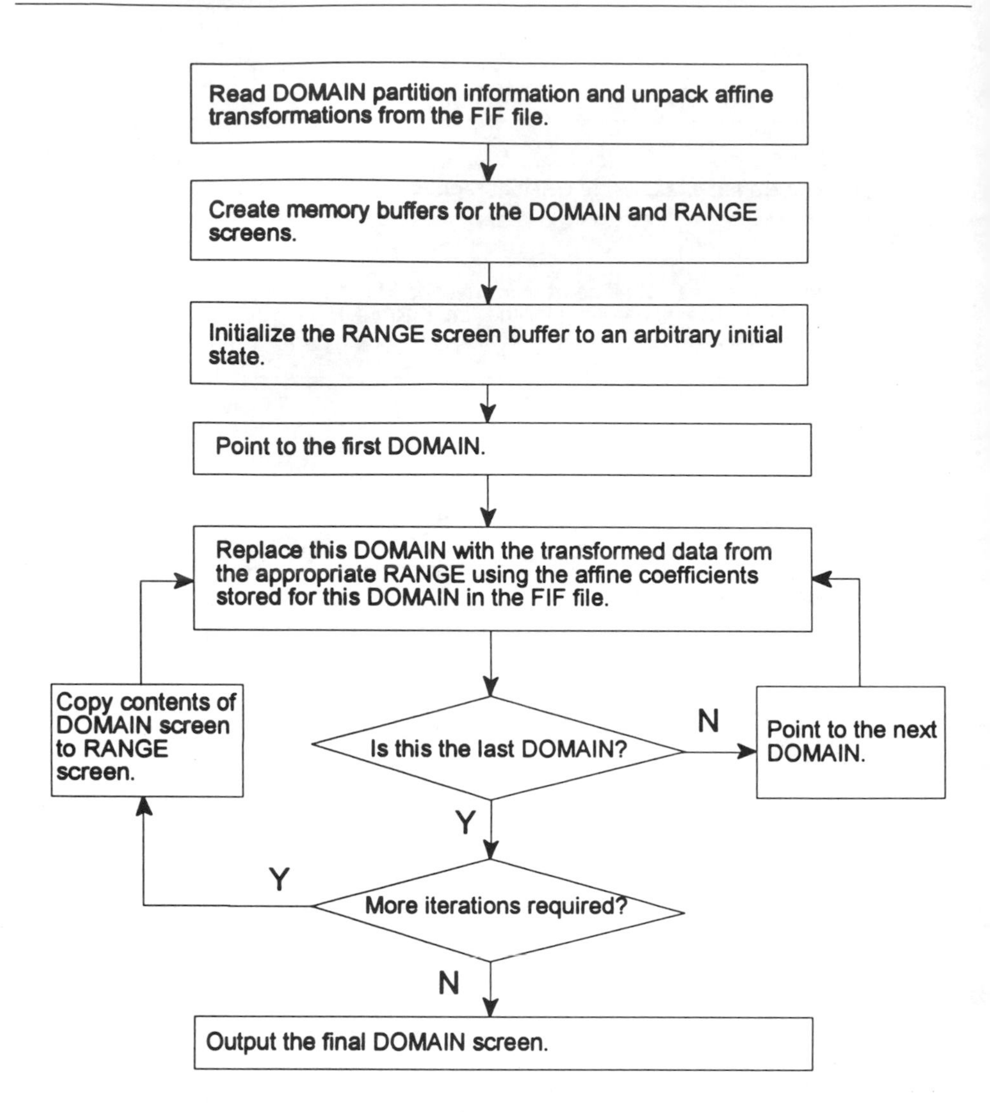

Figure 3.15 Fractal transform still-image decompression.

background; the areas where prediction failed are foreground. Those areas identified as background do not need to be transmitted since they are the same as the previous image within a tolerable error. The foreground areas are then fractal coded and transmitted.

3.6 OBJECT-BASED CODING TECHNIQUES

One difficulty with the H.261 coding algorithm is that blocking artifacts and mosquito noise are frequently generated at the boundary between moving objects and the static background. This is caused by the fact that the DCT block edges are not aligned with the edge of the object. *Object-based coding* (OBC) has been devised to attempt to minimize distortions of this type. The general principle of OBC [16–23] involves the identification of, and the encoding of, arbitrarily shaped objects moving within the scene. Information is transmitted about the object defining its motion, shape, and color. Two types of OBC systems have been investigated: (1) a generic approach dealing with unknown objects and (2) approaches dealing with known objects such as a talking head (also known as knowledge-based coding). The status of work in these two areas is outlined as follows.

3.6.1 Generic Unknown Objects

The University of Hanover, in Germany, has been a leader in the research of OBC. Scholars there have published many articles on the subject and have focused most of their energies on a specific implementation known as *object-based analysis-synthesis coding* (OBASC). Figure 3.16 is a functional block diagram of the OBASC encoder. The general architecture is the same as the H.261 coder in that they both employ interframe prediction and use the classic predictive loop structure. The parameter coder compares the current parameters with the stored/predicted parameters and transmits the prediction errors to the receiver. The loop is a little more complex than normal because not only are current/predicted parameters being compared, but current/synthesized images

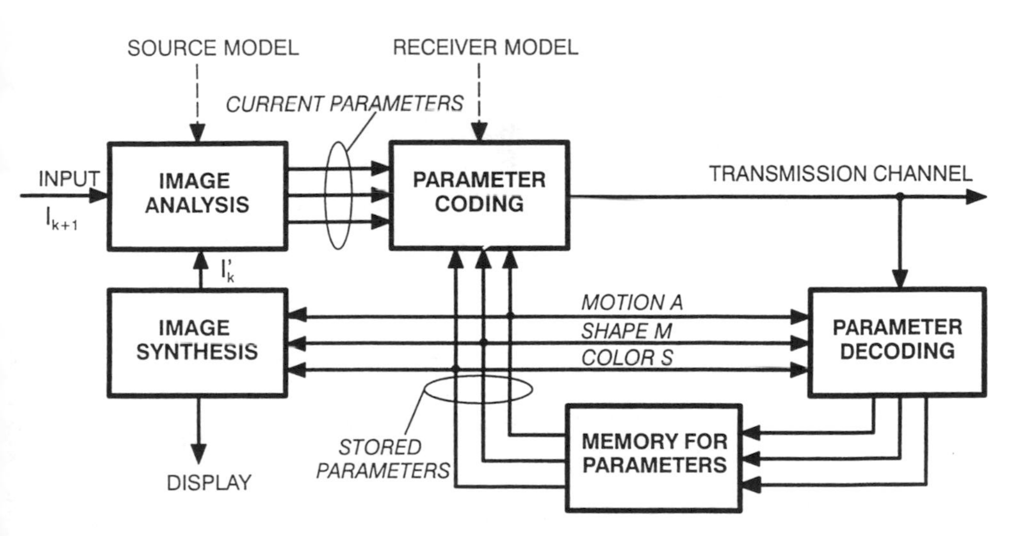

Figure 3.16 Block diagram of an object-based analysis-synthesis coder.

are also being compared by the image analyzer. The image analyzer develops the current parameter set.

An example of image analysis for the test sequence "Miss America" is illustrated in Figure 3.17. The analyzer decomposes the input images of a sequence into differently moving objects. For each object, three sets of parameters are determined describing the object's shape, motion, and color. In addition, each object is classified as to whether it complies with the underlying source model (that is, whether the changes of the color parameters can be described only by that object motion that is allowed by the source model) or whether the source model fails. Figure 3.17 illustrates the processing of the two classes of objects—model compliant and model failure.

A great deal of research is being directed toward source models having a wide range of complexity. The simplest source model restricts the definition of the object to a rigid two-dimensional format. More advanced models are capable of defining the objects as having 2D/flexible, 3D/rigid, and 3D/flexible structures.

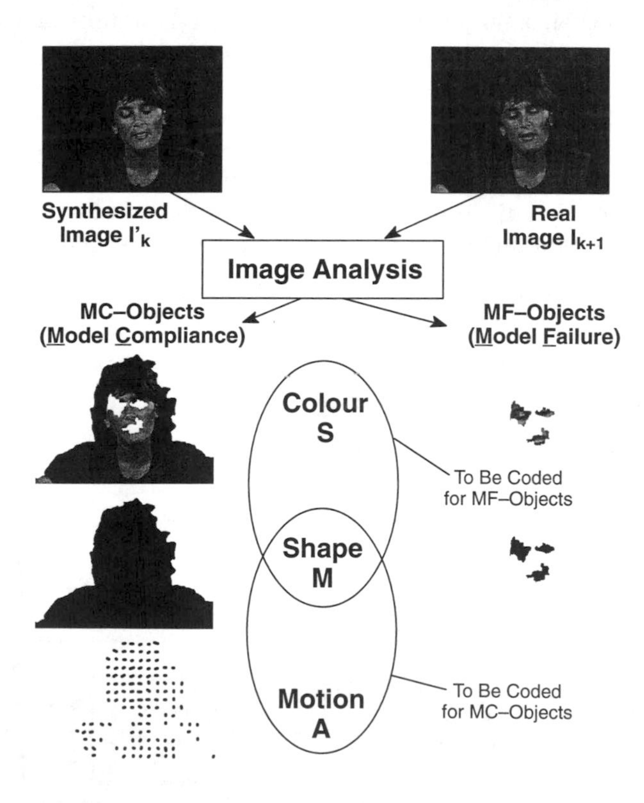

Figure 3.17 An example of image analysis output for the test sequence "Miss America."

Research is also under way on a variation of OBC known as region-based coding. In general, a region is an area having uniform brightness. Consequently, an object is usually made up of a number of regions. The advantage of region-based coding is that the requirements for defining the texture of the region has been eliminated. The disadvantage is that the number of arbitrarily shaped elements to be defined and transmitted has increased.

3.6.2 Knowledge-Based Coding

The OBASC system previously described is designed to detect and encode any generic objects that may appear in the scene. It does not take advantage of any apriori knowledge; for example, it may be very likely that some objects (such as head and shoulders) will occur very frequently. Knowledge-based coding systems are specifically designed to take advantage of this prior knowledge. The knowledge-based object is usually defined by a wireframe model as illustrated in Figure 3.18. The first step in the encoding process in to detect the existence of the known object in the scene and the adaptation of the wireframe model to the characteristics of the particular object. At this point the normal OBASC procedures of transmitting motion, shape, and color parameters take place.

Research is under way on a more advanced version of knowledge-based coding known as semantic coding. Semantic coding is applicable to a knowledge-based object that has a restricted set of action units. In the case of the head-and-shoulders object, a restricted set of action units could be (1) mouth open/closed or (2) eye open/closed. An example of semantic coding for the Miss America test scene is provided in Figure 3.19.

3.7 VARIABLE LENGTH CODING

VLC, also called entropy coding, is a technique whereby each event is assigned a code that may have a different number of bits. To obtain compression, short codes are assigned to frequently occurring events and long codes are assigned to infrequent events. The expectation is that the average code length will be less than the fixed code length that would otherwise be required. If all events are equally likely, or nearly so, then VLC will not provide compression.

All codes considered must be uniquely decodable; that is, there must be only one way that a concatenation of VLCs can be decoded. In addition, it is highly desirable that the code be instantaneous; that is, each code word can be decoded without reference to subsequent code words. Taken together, these requirements mean that no code word can be the beginning of another code word. For example, we may not have 01 and 0110 as code words, since the second code word starts with the first code word. In decoding, it is not known whether 01 is the first code word or just the start of the second code word.

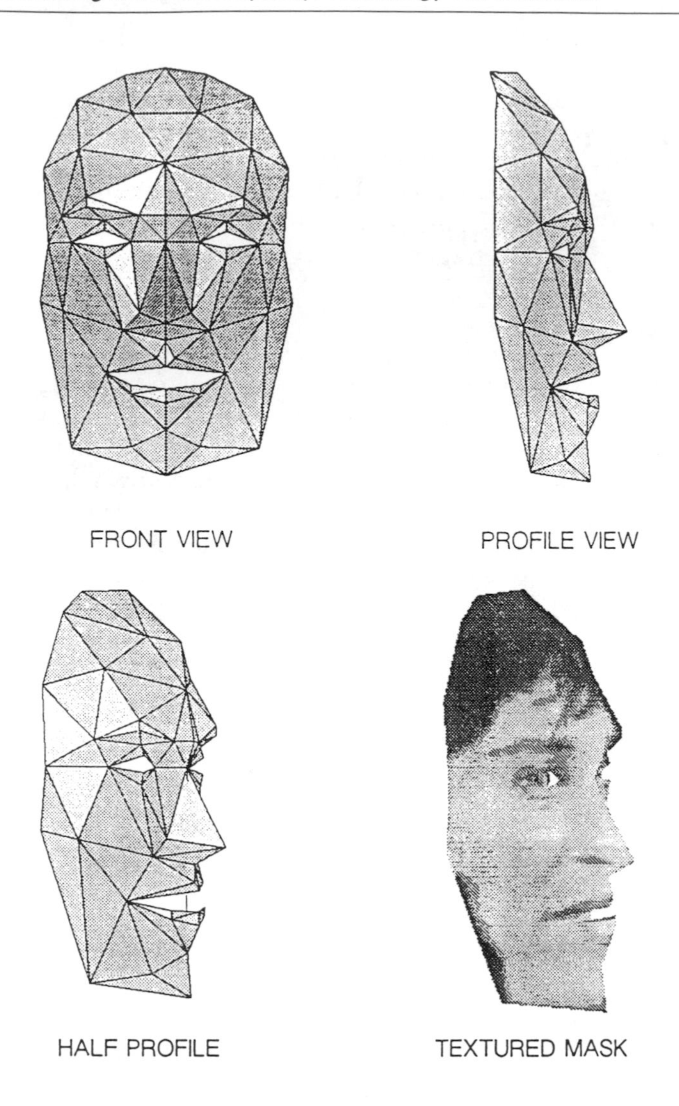

Figure 3.18 Model mask candide.

A major advantage of VLC is that it does not degrade the signal quality in any way. That is, the reconstituted signal will exactly match the input signal so that if the signal is adequately described by a series of events, using VLCs to communicate them to the decoder will not change the events. Therefore the system is transparent to the VLC used.

The disadvantage of VLCs is that they only provide compression in an average sense. Therefore, sometimes the code could be longer for a specific section of signal. This characteristic gives rise to the need for a buffer to match

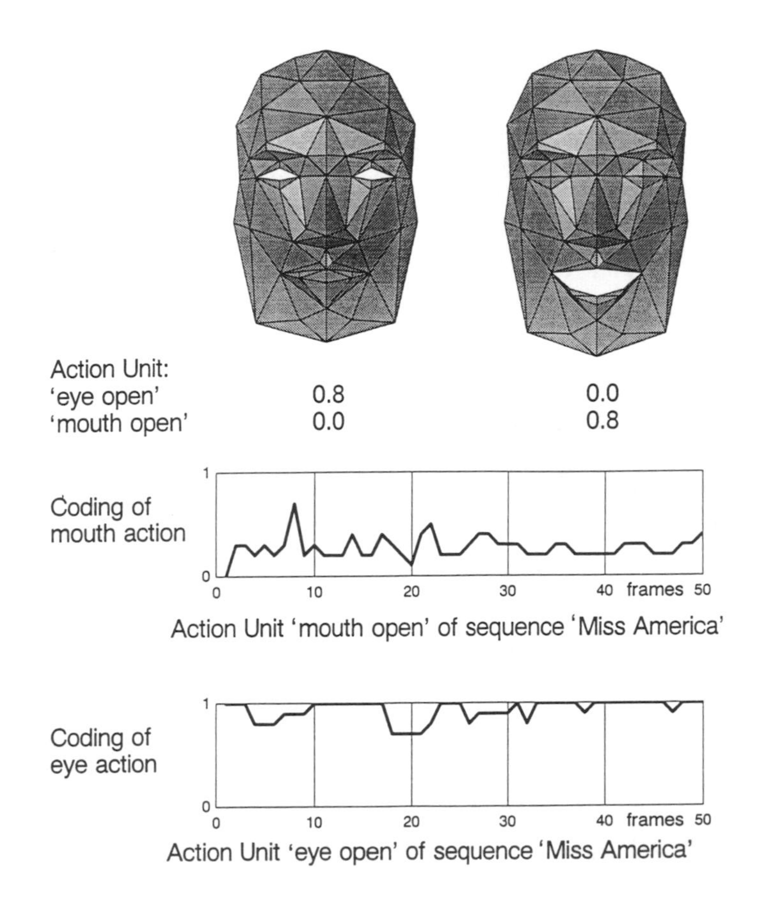

Figure 3.19 Example of semantic coding.

the variable rate of bit generation with the fixed bit rate of the communication channel and a control strategy to prevent long-term overflows or underflows of the buffer. Also the establishment of frames or packets of data becomes more difficult with VLCs.

Seven VLCs will be discussed here. They are comma codes, shift codes, B codes, Huffman codes, conditional codes, arithmetic codes, and two-dimensional codes.

3.7.1 Comma Code

The comma code is the simplest of the VLCs. It assigns to each event a different length of code, starting at 1. A particular bit polarity marks the end of the code word, such as the following.

Code
0
10
110
1110
11110
111110
.
.
.

The advantage of the comma code is that it is simple to generate and decode, requiring only counters to count the number of ones. However, it is rare that this code accurately matches the statistics of the events, so it is used primarily where simplicity of implementation is important.

3.7.2 Shift Code

In the case where the probabilities of the events decrease monotonically as the magnitudes increase, a great simplification can be obtained by using a systematic VLC, such as a shift code. In this code, each code word consists of a series of subwords, each of length L bits. The first subword is capable of conveying 2^L values, one of which is a shift that indicates that the value of the code word is contained in the following subword. In this way, any length of code word can be obtained by concatenating a number of subwords together.

The following are examples of the beginnings of shift code tables for $L = 1$, 2, and 3, where a subword of all 1s indicates a shift.

L = 1	L = 2	L = 3
0	00	000
10	01	001
110	10	010
1110	1100	011
11110	1101	100
111110	1110	101
1111110	111100	110
11111110	111101	111000
111111110	111110	111001
1111111110	11111100	111010
11111111110	11111101	111011
.	.	.
.	.	.
.	.	.
.	.	.

Note that for $L = 1$, the shift code reduces to the comma code. The shift code is best suited to cases where the probabilities drop off rapidly since the number of codes available only increases linearly with the length of the code. For example, for $L = 3$, there are seven codes with length 3 ($2^3 - 1 = 7$). Increasing the length to 6 only adds seven more codes.

3.7.3 B Code

Another variable length code that is systematic is the B code. Again the code consists of a sequence of subwords, each of length L. But in this case, one bit of the subword is used to designate whether another sub-word is to be added to the code word. Therefore the remaining L-1 bits in the subword can be used as part of the code. For $L = 1$, the B code also reduces to the comma code.

The following are examples of the beginning of B code tables for $L = 1$, 2, and 3.

L = 1	L = 2	L = 3
***********	* * * *	* * *
0	00	000
10	01	001
110	1000	010
1110	1001	011
11110	1100	100000
111110	1101	100001
1111110	101000	100010
11111110	101001	100011
111111110	101100	101000
1111111110	101101	101001
11111111110	111000	101010
.	111001	101011
.	111100	110000
.	111101	110001
	10101000	110010
	10101001	110011
	10101100	111000
	10101101	111001
	10111000	111010
	10111001	111011
	10111100	100100000
	.	.
	.	.
	.	.

In this table, the * marks the columns containing the continuation bits, where 1 indicates continue and 0 marks the last subword of the code word. This version of the B code is instantaneous. In another version, the continuation bit is the same value for all subwords in the code word but alternates with each succeeding code word. That version is not instantaneous.

The B code is best suited to cases where the probabilities drop off slowly since the number of codes available increases geometrically with increasing code length. For example, for $L = 4$, there are eight codes with length 4 (2^{L-1}). Increasing the length to eight increases the number of codes by 64 ($2^{2(L-1)}$).

3.7.4 Huffman Code

The Huffman code is a VLC that provides the shortest average code length for a given distribution of input probabilities. The method for generating the code is well-known, but a distribution of input probabilities, either theoretical or measured, is required before the code words can be calculated. If the actual distribution differs from that used to calculate the code, then the average code length may not be less than other codes. If a large enough sample can be obtained to measure the distribution accurately, the Huffman code may be an attractive choice. In any event, it provides a reference against which other codes can be compared, if the distribution is measured on the image being coded.

3.7.5 Conditional Variable Length Codes

In general, the most likely sample values to be encoded are near zero, and therefore the small values are given the shortest codes. However, zero is the most likely sample value only in the absence of information about other samples. If the values of neighboring samples are known, then the distribution of the current sample value can be markedly changed. In the simplest case, only the previous sample is used. Although, in principle, other samples can be used for each value of the previous sample, the frequency of occurrence of each of the current samples can be obtained, and a set of VLCs devised for each. Since both the encoder and decoder know the value of previous samples, decoding can take place without significant delay.

3.7.6 Arithmetic Coding

In arithmetic coding, the frequency of occurrence of the symbols to be coded is continuously measured by both the encoder and decoder. In the resulting code, there is not a one-to-one correspondence between the events and specific bits. It is possible to generate arithmetic codes at a rate of less than one bit per event, whereas a Huffman code requires at least one bit per event. Since arithmetic coders adapt dynamically to the statistics of the image being transmitted, the compression is generally superior to that for conventional nonadaptive VLCs.

3.7.7 Two-Dimensional VLC for Coding Transform Coefficients

A VLC has recently been developed that is particularly designed to code transform coefficients. It is a two-dimensional code where the two dimensions are the number of zero-value coefficients in a row (usually from a zig-zag scan) and the value of the next nonzero coefficient. This VLC is a likely code to be employed in the JPEG standard.

References

[1] "Component vector quantization," Annex 4 of CCITT Study Group VIII, Geneva, December 1–12, 1986.

[2] Abut, H., Ed., *Vector Quantization*, IEEE Acoustics, Speech, and Signal Processing Society, New York: IEEE Press, 1990.

[3] Gersho, A., "On the Structure of Vector Quantizers," *IEEE Trans. on Information Theory*, Vol. IT-28(2), March 1982, pp. 157–166.

[4] Gersho, A., and B. Ramamurthi, "Image Coding Using Vector Quantization," *Proc. ICASSP*, Paris, 1982.

[5] Gersho, A., and R. M. Gray, *Vector Quantization and Signal Compression*, Norwell, MA: Kluwer Academic Publishers, 1992.

[6] Gray, R. M., and Y. Linde, "Vector Quantizers and Predictive Quantizers for Gauss-Markov Sources," *IEEE Trans.on Communications*, Vol. COM-30(2), Feb. 1982, pp. 381–389.

[7] Helden, J., and D. E. Boekee, "Vector Quantization Using a Generalized Tree Search Algorithm," *Proc. 5th Symp. Information Theory in the Benelux,ri Aalten*, May 1984, pp. 21–27.

[8] Huang, H. M., and J. W. Woods, "Predictive Vector Quantization of Images," *IEEE Trans. on Communications*, Vol. COM-33(11), Nov. 1985, pp. 1208–1219.

[9] Linde, Y., A. Buzo, and R. M. Gray, "An Algorithm for Vector Quantizer Design," *IEEE Trans. on Communications*, Vol. COM-28(1), Jan. 1980, pp. 84–95.

[10] Akansu, A. N., and R. A. Haddad, *Multiresolution Signal Decomposition; Transforms, Subbands, and Wavelets*, New York: Academic Press, Inc., 1992.

[11] Chui, C. K., "An Introduction to Wavelets," *Wavelet Analysis and Its Applications*, Vol. 1, New Tork: Academic Press, Inc., 1992.

[12] Zhang, Y.-Q., and S. Zafar, "Motion-Compensated Wavelet Transform Coding for Color Video Compression," *IEEE Trans. on Circuits and Systems for Video Technology*, Vol. 2, Sept. 1992, pp. 285–296.

[13] Ohta, M., and S. Nogaki, "Hybrid Picture Coding with Wavelet Transform and Overlapped Motion-Compensated Interframe Prediction Coding," *IEEE Trans. on Signal Processing*, Vol. 41, Dec. 1993, pp. 3416–3424.

[14] Hilton, M. L., B. D. Jaweerth, and A. Sengupta, "Compression Still and Moving Images With Wavelets," *IEEE Trans. on Image Processing*, Vol. 2, No. 3, 1994.

[15] Chui, C. K., Ed., "*Wavelets: A Tutorial in Theory and Applications*," Wavelet Analysis and Its Applications, Vol. 2, New York: Academic Press, Inc., 1992.

[16] Hürtgen, B., and Th. Hain, "On the Convergence of Fractal Transforms," *Proc. IEEE Int. Conf. Acoustics Speech and Signal Processing ICASSP'94*, 1994. In print.
[17] Hürtgen, B., F. Mller, and C. Stiller, "Adaptive Fractal Coding of Still Pictures," *Proc. Int. Picture Coding Symp. PCS'93*, Lausanne, Switzerland, 1993.
[18] Jacquin, A. E., "A Novel Fractal Based Block-Coding Technique for Digital Images," *Proc. IEEE Int. Conf. Acoustics Speech and Signal Processing ICASSP'90*, Vol. 4, 1990, pp. 2225–2228, 1990.
[19] Jacquin, A. E., "Image Coding Based on a Fractal Theory of Iterated Contractive Image Transformations," *IEEE Trans. on Image Processing*, Vol. 1, No. 1, Jan. 1992, pp. 18–30.
[20] Lundheim, L., *Fractal Signal Modelling for Source Coding*, PhD thesis, Universitetet I Trondheim Norges Tekniske Høgskole, 1992.
[21] Aizawa, K., H. Harashima, and T. Saito, "Model-Based Analysis-Synthesis Image Coding (MBASIC) System for a Person's Face," *Signal Processing: Image Communication*, Vol. 1, No. 2, Oct. 1989, pp. 139–152.
[22] Hötter, M., and R. Thoma, "Image Segmentation Based on Object Oriented Mapping Parameter Estimation," *Signal Processing*, Vol. 15, No. 3, Oct. 1988, pp. 315–334.
[23] Hötter, M., "Optimization and Efficiency of an Object-Oriented Analysis-Synthesis Coder," *IEEE Trans. on Circuits and Systems for Video Technology*, Vol. 4, No. 2, April 1994, pp. 181–194.

4 Speech Coding

One important component of any audiovisual terminal for videoconferencing, videophone, and general multimedia applications is the speech coder. This chapter presents the attributes of speech coding and a review of currently available speech coders.

4.1 SPEECH CODER ATTRIBUTES

Speech coders have four attributes: bit rate, quality, complexity, and delay. For a given application some of these attributes are predetermined while tradeoffs can be made among the others. For example, quality can usually be improved by increasing bit rate or complexity and sometimes by increasing delay. In the following subsections, we discuss the various attributes with particular relevance to PSTN video telephony.

4.1.1 Bit Rate

Public-switched telephone networks (PSTN) video telephones are expected to operate at bit rates up to at least 28.8 Kbps. The higher the bit rate, the better the video quality. Given this fact, it is desirable that the speech coder use as little of the total bit rate as possible. Prior to Recommendations G.729 and G.723.1 ITU-T speech coding recommendations only existed for bit rates of 16 Kbps and higher. Rates lower than 9.6 Kbps have been used for digital cellular telephones, secure telephones, and satellite telephone services, such as Inmarsat provides.

4.1.2 Delay

Low-rate speech coders can be considered block coders. That is, they encode a block of speech, also known as a frame, at a time. Depending on the application, the total speech coding delay of the system is some multiple of the frame size. The minimum delay of the system is usually 3 or 4 times the frame size. In the case of low-bit rate speech coders, the total delays are far larger than those coders currently standardized in ITU-T G series recommendations. For example, many lower rate coders have frame sizes of 20 ms, resulting in a one-way delay of 60 ms to 80 ms. By contrast, 16 Kbps G.728 has a one-way delay of under 2 ms.

Even higher delays may result from the internal architecture of the overall videophone system. The delay of the video is usually much larger than that of the audio. For example, a video coder running at 5 frames per second has a frame size of 200 ms. The same 3 or 4 times the frame size rule also applies for the video coder. Thus, the minimum delay of the video is 600 ms to 800 ms in this case. The speech coder would have much less delay than the video coder. From experience with satellite links, we know that if the one-way speech coder delay is greater than 300 ms, users will notice and some will object. This is true even if there are no echoes. The echo problem is greatly exacerbated by large delays. Based on this evidence, all of the speech coders considered herein have reasonable delays. In fact, they might even be able to double their delay without having a perceived degradation in their quality. However, echo cancellers will be necessary because of the delay. As delay is increased, the inserted echo path loss requirements increase significantly. Therefore, there is a penalty in system design as delay is increased.

4.1.3 Complexity

Most speech coders are implemented first on *digital signal processing* (DSP) chips and then may later be implemented on special purpose VLSI devices. Speed and *random access memory* (RAM) are the two most important contributors to complexity. Generally one word of RAM takes up about as much space as six words of *read only memory* (ROM). In rating the complexity, the utilized scale used envisions the following types of 16-bit fixed point DSP chips. A low complexity DSP is capable of executing 10 to 20 *million instructions per second* (Mips) and has 512 to 1024 words of RAM and up to 6K words of ROM. A medium complexity DSP is capable of executing up to 30 Mips and might have up to 2K words of RAM and 10K words of ROM. A high complexity DSP chip is capable of 40 Mips or higher and might have more memory than the medium complexity part.

DSP chips also come in floating point. Speeds are typically slower, but more can be executed in a single instruction cycle. They usually require off-chip RAM, and the combination of external RAM and larger size results in larger power usage and greater overall system cost.

4.1.4 Quality

The very first digital speech coding standard is recommendation G.711 for 64 Kbps PCM speech. The distortion introduced by a G.711 codec is considered one *quantization distortion unit* (QDU). A *speech quality experts group* (SQEG) uses the QDU for network planning purposes. The second digital speech coding standard was G.721 32 Kbps ADPCM. This coder must be used in combination with a G.711 codec at its input and output. This combination is considered to have a distortion of 3.5 QDU. SQEG network planning guidelines call for a maximum of 14 QDU for an END-to-END international connection and less than 4 QDU for a domestic connection. The G.728 16 Kbps LD-CELP coder is considered to have the same QDU as G.721. These are the only coders considered toll quality.

Regional bodies have created standards for digital cellular telephony. The first generation of digital cellular standards all produce clear channel speech quality that is roughly comparable. Listeners can tell that the speech is not as clear as the original 64 Kbps coder, or as clear as G.721 or G.728. However, this level of quality is considered acceptable for cellular service and most users can hold a conversation for an extended period of time without listener fatigue due to the additional impairments. In 1995 several of the first generation cellular standards were upgraded to give better speech quality.

Lower bit rate speech coders have been created for other applications. In 1990 Inmarsat held a contest to determine a combined speech and channel coder for mobile satellite. The combined rate was 6.4 Kbps, but only about 4.15 Kbps was used for the speech coder. In 1989 work was done to create a U.S. federal standard for secure telephones operating at 4.8 Kbps. In both of these cases, the quality of the speech was less than IS-54 VSELP or RPE-LTP. The speech is still quite intelligible, but there is enough additional distortion that listeners have to strain to understand some words. This strain causes listener fatigue during a prolonged conversation. Previous ITU-T speech coders and those used in cellular standards have primarily been tuned for telephone handset use. In a videophone, they may also be used with a speakerphone. Low-bit rate speech coders can be tuned according to how they are used. A setting that sounds best for a handset may not sound that way for a loudspeaker. In addition, the microphone characteristics of a handset and a speakerphone are different. This means the input speech spectra for the two are different. Since a videophone coder could be used both ways, any coder selected should be tested for both modes and tuned to give the best overall compromise. In addition to clear channel quality, there is the issue of how a speech coder behaves on a channel having bit errors. In the case of digital cellular standards, provision is made for additional bits to be used for channel coding to protect the information bearing bits. Not all bits of a speech coder are equally sensitive. It is common practice to have two or three classes of bits with the most sensitive bits getting the most protection while the least sensitive class receives no protection at all. For low-bit rate applications such as secure telephones over 4.8 Kbps modems, it is reasonable to expect that the distribution of bit errors would be random. For high-rate modem signals, such as those used for video telephony, errors are more likely to occur in bursts due to the more complex modulation schemes used in these modems. In this case, the speech coder requires a mechanism to recover from an entire lost frame. This is referred to as a frame erasure concealment.

For the purpose of providing better video quality many videophones are considering the use of voice activity detectors with their speech coders. During nonspeech intervals the speech coder is turned off and the bits are allocated to the video coder to provide better video quality. At the receiver *comfort noise* is played out to simulate the background acoustic noise at the encoder. A similar method is used for some cellular systems and is also used in *digital speech interpolation* (DSI) systems. Most international phone calls carried on undersea cables or satellites use DSI systems.

There is some impact on quality when these techniques are used. Subjective testing can determine the degree of degradation. The methods typically suffer from two problems. If the voice activity detector is slow to detect speech, the onset of the speech sound is lost. This is referred to as front-end clipping. The performance of the voice activity detector is responsible for either the front-end clipping problem or its solution. The second problem occurs when there is an unusual acoustic noise in the background of the speech. The noise will interfere with the speech signal but will be eliminated during nonspeech intervals. This has a disconcerting effect on the listener and can impact intelligibility. Depending on the character of the background noise, unwarranted amounts of comfort noise may be inserted during the nonspeech intervals, resulting in the listener hearing more noise during nonspeech intervals.

4.1.5 Validation

Speech coding standards can be specified in different ways. At the very minimum, there are the bit stream standards. A good example of this is the LPC-10E vocoder used for secure voice terminals by NATO. At this point the standard is about 20 years old. Only the bit stream was ever specified. This meant that all of the parameters to be quantized were specified, together with their quantization tables and the order in which these parameters are to be transmitted. Over the years improvements have been made to both the encoder and the decoder to improve the quality and intelligibility of the output speech.

While this might seem to be a strength, it is also a weakness. It does not provide a customer with assurance that the hardware will have a specified level of quality. Over the years the governments purchasing this equipment have had to specify their own tests to determine if a vendor's equipment met their requirements. Since there were relatively few governments and contractors, this was still considered an adequate solution.

At the opposite end of the spectrum are bit exact specifications. Examples of such coders are G.721 (32 Kbps ADPCM) and the digital European cellular full rate standard (RPE-LTP). If digital speech samples are presented to a bit exact coder, then the coder will produce the same exact bit stream, regardless of the manufacturer of the coder. Bit exact standards have test vectors. A given input test vector must result in a given output test vector or a given bit stream. Any deviation will be considered a failure to conform to the specification. Consequently, the entire inner workings of the coder must be available to the implementer so that output test vectors and bit streams will match those specified. This also helps assure both customers and equipment providers that the terminal hardware will achieve the specified level of performance.

There are other possibilities that lie between these two. For G.728 (floating point), a mathematically exact specification was given. If implemented in floating point, most implementations were able to use commercially available DSP chips. Their outputs for test vectors were within a certain deviation of the official test vectors. The deviations were far below those detectable by the human ear. All concerned were assured that the coder's performance fully met all requirements.

For the North American TDMA digital cellular standard IS-54 the description of the coder is not as precise. A set of test vectors exists, but no one has exactly matched the output test vectors for a fixed point implementation. However, if their implementation is close enough to the output test vectors, then a manufacturer can be fairly certain that the coder is correctly implemented. The validation procedure for IS-54 also contains the description of a subjective test that can be given to make certain that the coder performs as specified.

The important point to be made here is that validation procedures need to exist. A sanctioned method for establishing conformance with the specification assures the manufacturer that the product can be implemented solely on the basis of the published specification. It assures the customer that the performance will be within specification. Finally, it assures network service providers that their service will meet the specified quality goals.

4.2 CURRENTLY AVAILABLE SPEECH CODERS

This section is divided into subsections on potential ITU coders for use in video teleconferencing. The three subsections contain information on speech coders that are ITU recommendations. All three can be used as part of ITU Recommendation H.320 for video conferencing. Only Recommendation G.729

can be used with ITU Recommendation H.324, as the other two coders have bit rates that are too high for ITU Recommendation H.324.

4.2.1 64-56-48 Kbps G.722 SBC

The only present standard for 7 kHz speech coding is ITU Recommendation G.722. Its principal applications are teleconferences and videoteleconferences. It is recommended for use with H.320 terminals on connections greater than 128 Kbps. The wider bandwidth (50–7000 Hz) is more natural sounding and less fatiguing than telephone bandwidth (200–3200 Hz). The wider bandwidth increases the intelligibility of the speech, especially for fricative sounds, like /f/ and /s/, which are difficult to distinguish for telephone bandwidth.

The G.722 coder is a two-band sub-band coder with ADPCM coding in both bands. It uses a 24 tap quadrature mirror filter to divide the signal into two bands. Each is ADPCM coded and transmitted. The ADPCM is similar in structure to that of ITU Recommendation G.727. The upper band uses an ADPCM coder with a 2-bit adaptive quantizer. The lower band uses an ADPCM coder with an embedded 4-5-6 bit adaptive quantizer. This makes bitrates of 48 Kbps, 56 Kbps, and 64 Kbps all possible. The 24 tap filter causes a delay of only 1.5 ms and is efficiently implemented using just 12 multiplications per decimated sample. With sample-by-sample ADPCM quantization in each band, 1.5 ms is also the delay of the coder. No voice activity detector is specified for this coder by the ITU.

The high fidelity of the lower band coding (up to 6 bits per sample) does a good job of masking the rather scratchy 2 bits per sample coding of the upper band. This example shows that some perceptual masking is possible between the telephone bandwidth and the higher frequencies. At the same time, it seems to indicate that the ear responds favorably to noticing the frequency content in the upper band, even if it is scratchy. In subjective testing comparisons with telephone bandwidth speech using binaural listening (i.e. with both ears), the 8 kHz bandwidth is always preferred, sometimes by as much as 1 MOS point.

Recommendation G.722 is a low complexity coder. It was designed to be implemented on the first generation DSP chips. It requires about 3 Mips and only a few hundred words of RAM. Since it was designed for the early fixed point DSP chips, it is specified in bit exact fixed point equations. Test vectors for G.722 are available from the ITU.

4.2.2 16-Kbps G.728 LD-CELP

The bit rate of this speech coder is 16 Kbps, which is too high for use in a very low bit rate videophone recommendation. However, it is commonly used in higher bit rate videoconferencing systems. The coder has other properties that set benchmarks for what would be desirable in any future standard. For example, its block size allows it to be used in H.320 with H.221 framing. In particular, its very first application is in ISDN H.320 videophones. So, it is known to be well-suited for videophones. One of the other potential applications for G.728 is digital circuit multiplication equipment that requires a variety of bit rates. Work is underway within ITU-T to modify G.728 to work at rates as low as 9.6 Kbps, but only the 16 Kbps version would have toll network quality.

This coder is considered to be toll quality. In combination with G.711 input and output, this coder is considered to be equivalent to 3.5 QDU or about 28-dB MNRU for *intermediate reference system* (IRS) weighted speech. The coder was extensively tested, including low-level and high-level inputs and was shown to behave similarly to other network quality standards such as G.721 32 Kbps ADPCM. The coder is remarkably robust to random bit errors. At 10^{-3} random bit errors, the coder is significantly more robust than any other ITU-T coder—the effects are barely perceptible. It also performs better at 10^{-2} BER than any other speech coder, but the effects are more noticeable. There was no ITU-T requirement for frame erasure concealment in the original terms of reference. Because of the interest in using this coder for Future Public Land Mobile Telephone Service, work is now under way to modify the decoder to provide frame erasure concealment.

The coder has been tested in more than a half-dozen languages and with various types of background noises and with music. No language dependency was found and no annoying artifacts were produced with nonspeech inputs. The coder has only been formally tested by the ITU-T over IRS microphone characteristics and telephone handsets for listening. It has been tested at AT&T with non-IRS flat-weighted speech and with headphone listening. It has always scored as well as the 32 Kbps ADPCM standard in these tests. It has also been played over loudspeakers many times. Many listeners have commented that it sounds better over a loudspeaker than 32 Kbps ADPCM.

Two tandem encodings are considered equivalent to 7 QDU. The coder was originally weakest in this area and revisions were made, which delayed the standard by one year. The result was the current performance, which actually exceeded the goals set in the terms of reference by matching the performance of 32 Kbps ADPCM for up to four tandem encodings.

The coder has approximately 2 ms one-way delay if implemented as described in the specification. It is possible to implement the coder in 2.5-ms blocks, which would result in a one-way delay of 7.5 ms to 10 ms, depending on other aspects of the processor and the system.

No voice activity detector is specified with the coder. As part of the workplan for the Future Public Land Mobile Telephone Service, it should be possible to create both a voice activity detector and a recommendation for operation of the coder with discontinuous transmission.

The floating-point recommendation for G.728 includes an appendix with implementation test vectors. Many laboratories have created C or Fortran language simulations that have passed the test vectors. DSP implementations based on standard floating-point DSP chips exist that have successfully passed the test vectors. A bit exact fixed point specification was created that is fully interoperational with any floating-point implementation that passes the floating-point test vectors.

G.728 is a high complexity algorithm. In floating point it takes 16 Mips to 18 Mips. A fixed point specification was finalized in 1994. The combined encoder and decoder have been implemented in as few as 27 Mips in fixed point using just under 2,048 words of RAM.

4.2.3 8 Kbps G.729 CS-ACELP

In response to a request from CCIR Task Group 8/1 for a speech coder for wireless networks as envisioned in the *Future Public Land Mobile Telecommunication Service* (FPLMTS), the ITU initiated a work program for a toll quality 8-Kbps speech coder in 1990. The original delay requirements were relaxed in 1991 to reduce the expected complexity of the coder. In 1993 and early 1994, two potential coders were tested. Both coders met the performance requirements for clean input speech quality. However, both failed the requirements for other input levels, for frame erasure concealment, and for speech with noisy backgrounds or music. In 1994 an agreement was reached that these coders would be combined and that their proponents would collaborate to produce an optimized coder capable of meeting all requirements.

In 1995 testing was repeated by four subjective test labs in four different languages. SQEG concluded that on the basis of these test results, the optimized coder met all speech quality requirements. Further characterization testing in late 1995 agreed with this conclusion. However, not all the tests were in agreement about the quality of speech with acoustic background noise. The more sensitive tests, such as the *comparison category rating* (CCR) test, indicated that there was lesser quality for G.729 than for the G.726 reference codec. However, it was the opinion of the Speech Quality Experts Group that the test methodology was not well enough understood to accept these results. *Absolute category rating* (ACR) test results contradicted the CCR results. The scores for G.729 were not significantly different than those of G.726 and were sometimes higher for noisy background conditions.

G.729 is a linear prediction analysis-by-synthesis coder. The whole name is Conjugate Structure Algebraic Codebook Excitation Linear Prediction. The conjugate structure is in the quantization of the linear prediction coefficients.

The algebraic codebook refers to the partitioning of the excitation space. This allows for a fast, almost optimal search of all possible excitation sequences. Two sub-frames of 5 ms each are combined with a single estimate of the pitch period and gain to form the excitation. The LPC filter is updated every 10 ms and is quantized in the line spectral pair domain using the conjugate quantization.

G.729 has a 10 ms frame size and 5 ms look-ahead. This results in a 35 ms one-way delay when computation, algorithmic delay and transmission are included. The complexity is reported to be about 15 Mips and about 3,300 words of RAM are required. This coder and G.723.1 are the first ITU coders to be specified by a bit exact fixed point ANSI C code simulation of the encoder and decoder. This recommendation was put forward for determination by SG15 in November 1995 and became an official ITU recommendation in 1996. G.729 is also the basis for the ITU's simultaneous voice and data speech coder. Annex A of Rec. G.729 is scheduled for approval in May 1996. It has a reduced complexity search. The coder's overall complexity is reduced to under 10 Mips, while still using the same bit stream format as G.729.

5 Audiovisual Standards Organizations

It has been clearly established that standards are a fundamental necessary step in the provision of effective audiovisual communications. The Group 3 facsimile and H.320 videoconferencing standards are excellent examples. Figure 5.1 illustrates the primary international and U.S. organizations that establish these standards. Section 5.1 focuses on the international standard organizations such as the ITU, ISO, and IMTC. It should be noted that the ITU and ISO organizations are fundamental standard-setting structures in that the basic membership is comprised of countries. In contrast, the IMTC is a consortium of corporations. Two key U.S. standard organizations—T1 and TIA—are discussed in Section 5.2.

5.1 INTERNATIONAL STANDARDS ORGANIZATIONS

5.1.1 International Telecommunications Union (ITU)

5.1.1.1 Mission, Structure, and Operation of the ITU

The ITU is a global intergovernmental treaty organization within which governments and the private sector cooperate to develop recommendations that guide telecommunication service providers and manufacturers of related equipment to provide interoperable telecommunication services. The ITU also allocates radio frequency spectrum and supports efforts to eliminate harmful interference between radio stations of different specific technical assistance projects funded by the United Nations. These general task outlines are pursued by the ITU via a three-sector organizational structure reflected in Figure 5.2 entitled "Organizational Structure of the ITU." Each sector is supported by an ITU Bureau headed by a Director with general administration managed by a General Secretariat all under the Secretary-General of the ITU.

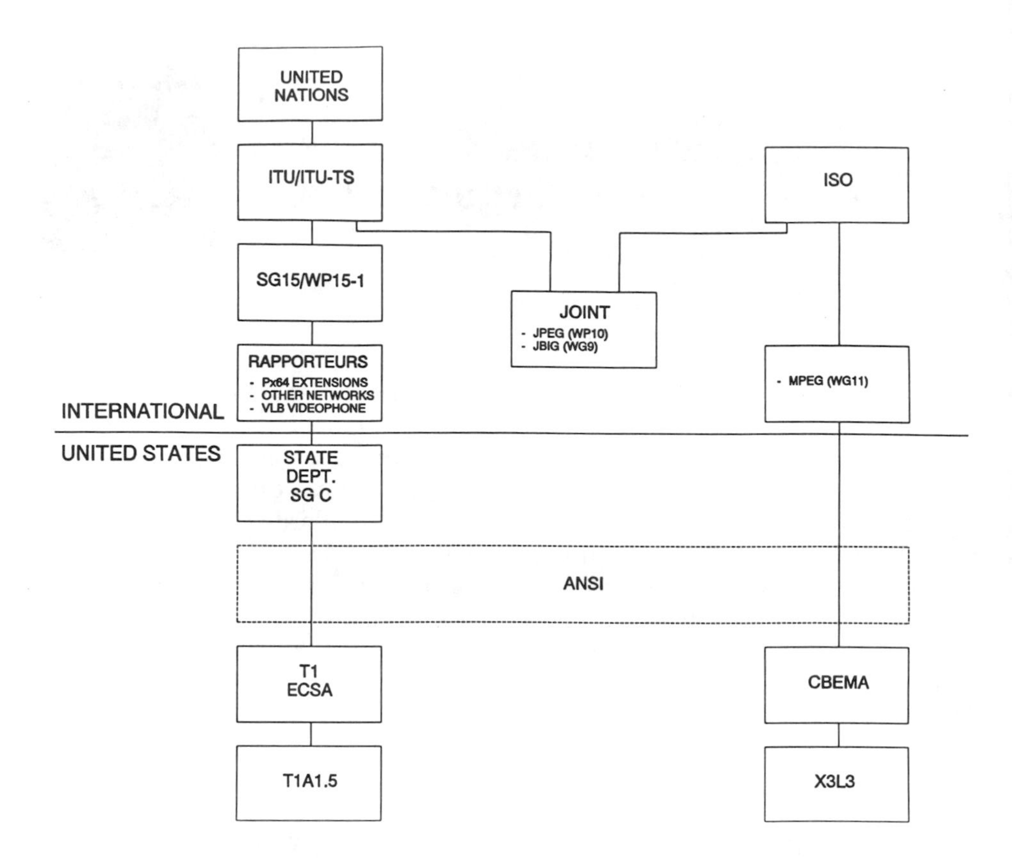

Figure 5.1 Organizations contributing to digital TV standards.

The method of operation in the ITU-TS is by way of *consensus*. The concept of consensus should be understood in terms of its practice in the ITU. The view most often supported is that consensus is arrived at when no Member objects to the proposal at hand or when only one or two Members wish to indicate their objections to some aspect of the proposal but do not resist it as a whole. The most difficult scenario is the one in which one or two Members object to the proposal in its entirety. In such case, chairman decision and voting are not appropriate mechanisms at Working Party or Study Group meetings except when the recommendation approval process is in progress at a Study Group meeting. No one suggests that ITU consensus is the same as T1 consensus or consensus as it might be defined elsewhere. The success of the ITU to date may be at least partially attributed to the use of consensus as practiced therein.

A few years ago, the ITU was organized on a rigid four-year work period such that standards were established every four years. They were published in

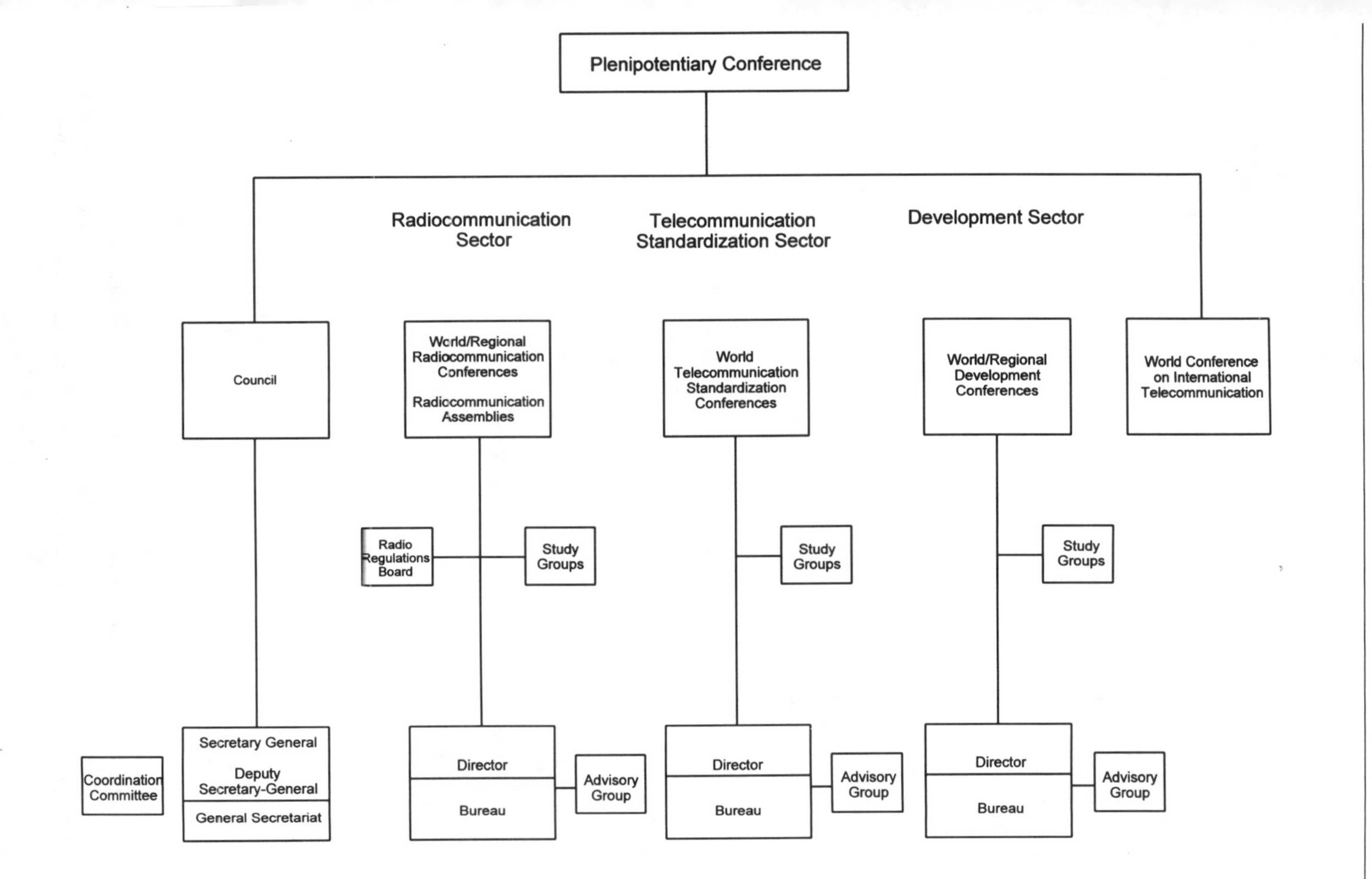

Figure 5.2 Organizational structure of the ITU.

sets of, for example, "Red" books and "Blue" books. It was determined that this process was too slow, and the Resolution 1 process was instituted that permits standards to be established at any time. The Resolution 1 process specifies that a recommendation is established using a three-step sequence. First, a Study Group "determines" that a draft recommendation is sufficiently stable to give notice of the intention to invoke the Resolution 1 process. At a subsequent meeting, the Study Group takes the second step of "deciding" to invoke the Resolution 1 process. Finally the draft recommendation is submitted to membership countries for ballot.

5.1.1.2 Telecommunication Standardization Sector

The *Telecommunication Standardization Sector* (ITU-T) was created on March 1, 1993 within the framework of the "new" ITU, along with the *Radiocommunications Sector* (ITU-R) and the *Telecommunication Development Sector* (ITU-D). The ITU-T replaces the former CCITT.

The purpose of the ITU-T is the elaboration, adoption, distribution, and follow-up of recommendations (nonbinding standards) to standardize telecommunications on a worldwide basis. These recommendations are developed from the study of Questions (that is, areas for study) for international telecommunications, in 15 Study Groups, by some 4,000 experts from throughout the world.

The standardization process is cyclic; experts elaborate draft standard texts, discuss them during periodic Study Group meetings, amend them after discussion, present revised text to the next meeting, and so on, until the Study Group considers the text to be mature enough to be presented to the Members (Administrations) or ITU for approval as an ITU-T recommendation. This approval is done in most cases by correspondence.

By right, all ITU Members may participate in ITU-T. Other entities and organizations, with the approval of the Administrations, can become "members" (with small "m") and participate also.

Such entities and organizations could include:

- Network or service providers, manufacturers or suppliers, financial or development institutions;
- Regional organizations and other international telecommunication, standardization, financial or development organizations;
- Other entities dealing with telecommunication matters.

As indicated earlier, the high-level ITU organization responsible for setting standards, in the videoconferencing and videophone areas, is the ITU-T. The ITU-T in turn is divided into fifteen Study Groups, which are listed in Table 5.1. The work performed by the ITU is defined in terms of "Questions"

Table 5.1
Telecommunication Standardization Sector (ITU-T)

Study Group	*Function*	*Recommendations*
1	Service definition	
2	Network operation	
3	Tariff and accounting principles	
4	Network maintenance	
5	Protection against electromagnetic environmental effects	
6	Outside plant	
7	Network operation	
8	Terminals for telematic services	T.120 Series, Fax
9	Television and sound transmission	
10	Languages for telecommunication applications	
11	Switching and signaling	
12	End-to-end transmission performance of networks and terminals	
13	General network aspects	I-SERIES (ISDN)
14	Modems and transmission techniques for data, telegraph, and telematic services	V.8, V.34, V.32
15	Transmission systems and equipment	H.320, H.324, H.310, H.321, G.SERIES (Speech)

that define the standards to be developed. The ITU assigns a "Rapporteur" to be responsible for the work performed for each of the Questions, and each Rapporteur assembles a group of experts to perform the work. The work is performed by experts submitting technical contributions for review by the Experts Group and by the supervisory Study Group.

The ITU-T structure comprises:

- The *World Telecommunication Standardization Conference* (WTSC), which defines general policy for the Sector, approves the program of work through the Questions for study, and allocates them to the Study Groups;
- The *Telecommunication Standardization Advisory Group* (TSAG), which looks at work priorities and strategies for the Sector, follows up on accomplishments of the work program, and advises the TSB Director and the Study Groups;
- The telecommunication standardization *Study Groups* (SGs) and their *Working Parties* (WPs), which bring together the experts who study the Questions and elaborate the Recommendations;

- The *Telecommunication Standardization Bureau* (TSB), which provides the Sector secretariat and coordinates the standardization process.

5.1.1.3 World Telecommunication Standardization Conference

The *World Telecommunication Standardization Conference* (WTSC) normally meets every four years to examine matters that are specific to telecommunication standardization, adopted by established procedures, or are posed by the Plenipotentiary Conference (supreme organ of ITU) or by the Council (executive organ of the Plenipotentiary Conference).

The WTSC is responsible for the work program, working methods, and structure of the Study Groups and the procedures for approving Recommendations. It:

- Examines Study Group reports;
- Establishes the work program resulting from examining the Questions, according to their urgency and priority, and evaluates financial impact and establishes a schedule to complete the corresponding studies;
- Decides whether to maintain or dissolve existing Study Groups or to create new ones and fixes their work program;
- Supervises the activities of the TSB, examines and approves the Director's report on Sector activities since the preceding conference;
- Approves, modifies, or rejects recommendations contained in the reports, although most of them are approved by correspondence.

The first WTSC called within the new ITU framework took place in Helsinki in 1993. It provided the Questions for study by the appropriate Study Groups during the new Study Period (the period between two successive conferences), that is, for the years 1993 to 1996. The next WTSC is planned for October 1996.

5.1.1.4 Telecommunication Standardization Advisory Group

Due to the constant and rapid evolution of telecommunications, the WTSC constituted the *Telecommunication Standardization Advisory Group* (TSAG) to:

- Study the priorities and strategies to adopt in the framework of ITU-T activities;
- Examine the progress in the execution of the work program for the Sector;
- Formulate recommendations with regard to the work of the Study Groups;

- Recommend measures to help the cooperation and coordination between ITU-T, the other ITU Sectors, and the ITU Strategic Planning Unit.

The TSAG follows the work of ITU-T, identifies changing requirements, and provides advice on appropriate changes, giving due regard to the cost and availability of resources. The TSAG also analyses Sector working methods and proposes possible ways of improving these methods.

To fulfill its mission, the TSAG has working groups to deal with:

- Computerized work program/priorities for Study Groups;
- Work programs external to ITU and projected results, so as to harmonize the ITU-T activities with those of other standardization organization;
- Follow-up on work progress within Study Groups; Promotion of electronic document handling.

5.1.1.5 Telecommunication Standardization Bureau

A Director, elected by the ITU Plenipotentiary conference, directs the TSB and organizes and coordinates the activities of the ITU-T.

The Bureau:

- Provides the necessary administrative, logistic and technical support to World Conferences and to telecommunication standardization Study Groups and their Working Parties;
- Applies the provisions of the International Telecommunication Regulations;
- Coordinates the elaboration, editing, and distribution of circulars, documents and publications for the Sector;
- Facilitates the exchange of administrative, operational, statistical and tariff information among the agencies operating public international telecommunication services, notably through the ITU Operational Bulletin;
- Provides technical information on international telecommunications and collaborates closely with both the Telecommunication Development Sector, for matters of interest to developing countries, and the Radiocommunications Sector.

The ITU's Telecommunication Standardization Bureau can be reached at the following addresses and numbers:

Street address: CH 1211 Geneva 20, Switzerland
Telephone: +41 22 730 5852
Facsimile: +41 22 730 5853
Telex: 421 000 uit ch
Internet email:tsbmail@itu.ch
X.400email:s=tsbmail; P=itu; A=arcom; C=ch

5.1.2 International Standards Organization (ISO)

The *International Standards Organization* (ISO) is a worldwide federation of national standards bodies from some 100 countries, one from each country. ISO is a nongovernmental organization established in 1947. The mission of the ISO is to promote the development of standardization and related activities in the world with a view to facilitating the international exchange of goods and services and to developing cooperation in the spheres of intellectual, scientific, technological, and economic activity. ISO's work results in international agreements, which are published as international standards.

5.1.2.1 How It All Started

International standardization began in the electrotechnical field; the *International Electrotechnical Commission* (IEC) was created in 1906. Following a meeting in London in 1946, delegates from 25 countries decided to create a new international organization "the object of which would be to facilitate the international coordination and unification of industrial standards." The new organization, ISO, began to function officially on 23 February 1947.

5.1.2.2 International Standardization : What Does It Achieve?

Industrywide standardization is a condition existing within a particular industrial sector when the large majority of products or services conform to the same standards. It results from consensus agreements reached between all economic players in that industrial sector, such as suppliers, users, and often governments. They agree on specifications and criteria to be applied consistently in the choice and classification of materials, the manufacture of products, and the provision of services. The aim is to facilitate trade, exchange, and technology transfer through:

- Enhanced product quality and reliability at a reasonable price;
- Improved health, safety and environmental protection, and reduction of waste;

- Greater compatibility and interoperability of goods and services;
- Simplification for improved usability;
- Reduction in the number of models and thus reduction in costs;
- Increased distribution efficiency and ease of maintenance.

Users have more confidence in products and services that conform to international standards. Assurance of conformity can be provided by manufacturers' declarations or by audits carried out by independent bodies.

5.1.2.3 Why Is International Standardization Needed?

The existence of nonharmonized standards for similar technologies in different countries or regions can contribute to so-called technical barriers to trade. Export-minded industries have long sensed the need to agree on world standards to help rationalize the international trading process. This was the origin of the establishment of ISO.

International standardization is now well-established for very many technologies in such diverse fields as information processing and communications, textiles, packaging, distribution of goods, energy production and utilization, shipbuilding, banking, and financial services. It will continue to grow in importance for all sectors of industrial activity for the foreseeable future.

5.1.2.4 Who Makes Up ISO?

A *member body of ISO* is the national body "most representative of standardization in its country." It follows that only one such *body for each country* is accepted for membership. The member bodies have four principal tasks, namely, to:

- Inform potentially interested parties in their country of relevant international standardization opportunities and initiatives;
- Organize so that a concerted view of the country's interests is presented during international negotiations leading to standards agreements;
- Ensure that a secretariat is provided for those ISO technical committees and subcommittees in which the country has an interest;
- Provide their country's share of financial support for the central operations of ISO, through payment of membership dues.

5.1.2.5 Who Does the Work?

The technical work of ISO is highly decentralized, carried out in a hierarchy of some 2,700 technical committees, subcommittees and working groups. In these committees, qualified representatives of industry, research institutes, government authorities, consumer bodies, and international organizations from all over the world come together as equal partners in the resolution of global standardization problems.

The major responsibility for administrating a standards committee is accepted by one of the national standards bodies that make up the ISO membership (for example, AFNOR, ANSI, BSI, CSBTS, DIN, and SIS). The member body holding the secretariat of standards committee normally appoints one or two persons to conduct the technical and administrative work. A committee chairman assists committee members in reaching consensus. Generally, consensus will mean that a particular solution to the problem at hand is the best possible one for international application at that time.

The Central Secretariat in Geneva acts to ensure the flow of documentation in all directions, to clarify technical points with secretariats and chairmen, and to ensure that the agreements approved by the technical committees are edited, printed, submitted as draft international standards to ISO member bodies for voting, and published. Meetings of technical committees and subcommittees are convened by the Central Secretariat, who coordinates all such meetings with the committee secretariats before setting the date and place. Although the greater part of the ISO technical work is performed via correspondence, there are, on average, a dozen ISO meetings taking place somewhere in the world every working day of the year. Each member body interested in a subject has the right to be represented on a committee. International organizations, governmental and nongovernmental, in liaison with ISO, also take part in the work. ISO collaborates closely with the IEC on all matters of electrotechnical standardization.

The scope of ISO is not limited to any particular branch; it covers *all standardization fields* except electrical and electronic engineering, which is the responsibility of IEC. The work in the field of information technology is carried out by a joint ISO /IEC technical committee (particularly JTC 1).

5.1.2.6 How Are ISO Standards Developed?

ISO standards are developed according to the following principles.

- *Consensus:* The views of all interests—such as manufacturers, vendors and users, consumer groups, testing laboratories, governments, engineering professions, and research organizations–are taken into account.
- *Industrywide:* Global solutions are sought to satisfy industries and customers worldwide.
- *Voluntary:* International standardization is market-driven and therefore based on voluntary involvement of all interests in the market price.

There are three main phases in the ISO standards development process.

The need for a standard is usually expressed by an industry sector, which communicates this need to a national member body. The latter proposes the new work item to ISO as a whole. Once the need for an international standard has been recognized and formally agreed upon, the first phase involves defining the technical scope of the future standard. This phase is usually carried out in working groups that are comprised of technical experts from countries interested in the subject matter.

Once agreement has been reached on which technical aspects are to be covered in the standard, a second phase is entered during which countries negotiate the detailed specifications within the standard. This is the consensus-building phase.

The final phase comprises the formal approval of the resulting draft international standard (the acceptance criteria stipulate approval by two-thirds of the ISO members that have participated actively in the standards development process and approval by 75% of all members that vote), following which the agreed text is published as an ISO international standard.

5.1.2.7 Searching for Information

Enquiries about standards involve those of ISO and a number of recognized standards agreed within other international technical organizations. There are, in addition, several hundred thousand standards and technical regulations in use throughout the world containing special requirements for a particular country or region. Finding information about all these standards, technical regulations, or related testing and certification activities can be a heavy task.

The *ISO Information Network* (ISONET) is there to ease the problem. This is a worldwide network of national standards information centers that have cooperatively developed a system to provide rapid access to information about standards, technical regulations, and testing and certification activities currently used in different parts of the world. Members of this network—usually the ISO member for any given country—act effectively as "specialized enquiry points" in the dissemination of information and in identifying the relevant sources of information for solving specific problems. The Agreement on Technical Barriers to Trade, drawn up under the *General Agreement on Tariffs and Trade* (GATT) (that is, the future *World Trade Organization* (WTO)), calls upon its signatories to establish in each country an enquiry point capable of answering questions about the standards, technical regulations, and certification systems in force in that country. In many countries, the ISONET enquiry point and the GATT enquiry point are one and the same.

5.1.3 International Multimedia Teleconferencing Consortium (IMTC)

The *International Multimedia Teleconferencing Consortium, Inc.* (IMTC) is a nonprofit corporation founded to promote the creation and adoption of international standards for Multipoint Document and Video Teleconferencing, specifically the ITU T.120 and H.320 standards suites. The corporation was formed in September 1994 through a merger of two other organizations, specifically the Consortium for Audiographics Teleconferencing Standards, Inc. (CATS) and the Multimedia Communications Community of Interest (MCCOI). CATS, formed in 1993, had focused on the ITU T.120 standards suite for Audiographics Teleconferencing; MCCOI, also formed in 1993, had focused more on the ITU H.320 standards suite for Video Telephony. The IMTC therefore brings together a large number of organizations at the cutting edge of these technologies worldwide. Through an active worldwide membership and corresponding activities, the corporation expects to provide strong impetus to the emerging multimedia teleconferencing market globally.

The IMTC, which is headquartered in San Ramon, California, is composed entirely of members from the United States, Canada, Europe, and Asia. The corporation's structure includes a board of directors, which oversees the management of the corporation and establishes its operating policies; and a number of activity groups, where much of the corporation's day-to-day activities take place. While the IMTC members do not make a specific time commitment when they join, the corporation does encourage full, active, and open participation from all its members.

5.1.3.1 Toward a New Communications Paradigm

The IMTC bases its existence on the fundamental changes experienced by the business community in recent years. These changes have two primary characteristics, particularly, globalization of markets and products, and in many cases, a shrinking pool of human resources in many corporations to perform the added work.

As these changes have gathered momentum, one fact has become evident: Today, more than ever before, individuals across the globe require new ways to communicate—in effect, a new communications paradigm. This involves the exchange of audio, video, graphics, and still image information in real time by individuals located at multiple sites anywhere in the world. Used together or in any combination, these technologies allow people to meet electronically, share information, solve problems, reach decisions, and at the end of the session move on to the project without leaving their corporate conference room or desktop.

This new communications technique, called multipoint multimedia teleconferencing (also known as data and/or document conferencing), actually embodies voice, graphics, video, and data. Judging by the number of companies involved in developing related products and services and the reaction of end users, it has the potential to meet the challenge of the 1990s, that is, the ability for corporations and individuals to optimize productivity without significant burnout of the steadily dwindling resources available in their organizations.

5.1.3.2 IMTC Objectives

While the applications of multimedia teleconferencing are tangible and the benefits easily quantified, it is clear that one hurdle must be crossed to ensure ubiquity. End users consistently point to the need for standards-based solutions to ensure that they are not tied to any one supplier's proprietary technology or limited to working in these restricted environments. Their goal is to protect and leverage corporate capital investment in new technologies, and this requires products manufactured by many different suppliers to interoperate, or interwork, with each other. In turn, this establishes a clear need for international standards, for which suppliers can adhere upon adoption.

The IMTC has a firm understanding of this perspective. The corporation's fundamental goal is to bring all organizations involved in the development of multimedia teleconferencing products and services together to help create and promote the adoption of the required standards. The IMTC's specific objectives are as follows.

- To promote and facilitate the broad use of multimedia teleconferencing based on open standards, including the H. 320 and T. 120 series of standards adopted by the ITU. The IMTC intends to consider similar and consistent standards from the ITU, other international standards bodies, open industry consortia, and organizations that utilize techniques that enjoy broad commercial acceptance, providing they do not conflict with the objective of promoting the growth of effective, open and widely available multimedia communications services and products.
- To educate the general business and consumer communities as to the value and benefits of open standards-based multimedia collaborative work through public statements, publications, trade show demonstrations, and other programs established by the corporation.
- To provide a forum for the organization's individual members to meet and cooperate on the definition of implementation specifications to promote full interworking of multimedia teleconferencing applications, products, and services.

- To encourage its members to make appropriate submissions to established communications agencies and standards bodies in support of open, de-jure standards.
- To support the testing for interworking and interoperability of multimedia teleconferencing products including applications, *customer premises equipment* (CPE), networking equipment, and services that support the global interworking of desktop, multimedia, collaborative solutions compliant with H.320, and T.120 teleconferencing standards as well as other appropriate open standards and specifications.

5.1.3.3 Benefits

By combining their resources to work on common issues, the IMTC members gain a number of benefits, including:

- Early input and access to T. 120 and H.320 standards data and direction;
- Ability to bring T.120 and H.320 compliant products to market earlier;
- Knowledge of products being developed by other members of the IMTC;
- Assurance that products are in line with industry trends;
- Return on investment in technology and products.

5.1.3.4 Membership

Membership in the IMTC is open to any and all interested parties, including teleconferencing hardware and software vendors, telecommunications companies, teleconferencing service providers, end users, educational institutes, government agencies, and nonprofit corporations.

Voting members participate in all the IMTC activities and functions and have full voting rights on all issues. The cost of a voting membership is $5,000 (U.S.) a year.

Non-profit organizations and educational institutes participate in all the IMTC activities and functions but may not vote. The cost of membership is $1,000 (U.S.) a year.

5.1.3.5 Committees

The IMTC currently has six activity groups that are responsible for the following activities.

- *API and Protocols:* This group is responsible for the recommendation of an Application Programming Interface and the communications protocols required in the T.120 and H.320 suites.

- *Interoperability Trials:* This group is responsible for all trials, such as H.320 and T.120, related to platform and application interoperability and interworking.
- *H.320 in the Local Area:* This group is responsible for analyzing and specifying the techniques necessary to transport full multimedia calls to the desktop in LAN environments.
- *Multipoint/Multiparty:* This group is responsible for determining technical and other requirements of multipoint and multiparty teleconferences.
- *Customer Requirements:* This group is responsible for determining actual end-user and market requirements and inputting this information to the other activity groups.
- *External Relations:* This group is responsible for marketing the IMTC message to the media, end-users, and other communities of interest.

5.2 U.S. STANDARDS ORGANIZATIONS

5.2.1 Standards Committee T1-Telecommunications

5.2.1.1 Introduction

Incorporated as a not-for-profit association in 1983, the *Exchange Carriers Standards Association* (ECSA) became the *Alliance for Telecommunications Industry Solutions* (ATIS) in 1993. ATIS consists of and is supported by members of the telecommunications industry to address exchange access, interconnection, and other technical issues in a postdivestiture telecommunications era. ATIS sponsors and provides secretariat support to Committee T1—Telecommunications Committee accredited by the *American National Standards Institute* (ANSI). ATIS sponsors a variety of postdivestiture industry forums established to discuss and resolve problems between and among, for example, exchange carriers, interexchange carriers, and enhanced-service providers, concerning communications services.

Established in February 1984, Committee T1 develops technical standards and reports regarding interconnection and interoperability of telecommunications networks at interfaces with end-user systems, carriers, information and enhanced-service providers, and CPE. Committee T1 has six technical subcommittees that are advised and managed by the *T1 Advisory Group* (T1AG). Each technical subcommittee develops draft standards and technical reports in its designated areas of expertise. The subcommittees recommend positions on matters under consideration by other national and international standards bodies. Technical subcommittees and their areas of expertise are:

- T1A1: performance and signal processing;
- T1E1: interfaces, power, and protection of networks;

- T1M1: internetwork operations, administration, maintenance, and provisioning;
- T1P1: systems engineering, standards planning, and program management;
- T1S1: services, architectures, and signaling;
- T1X1: digital hierarchy and synchronization.

Membership and participation in Committee T1 are open to all parties with a direct and material interest in the T1 process and activities. Free of dominance by any single interest, T1's policy of open membership and balanced participation safeguards the integrity and efficiency of the standards formulation process. ANSI's due process procedures further ensure fairness. Required procedures include announcing meetings in advance, distributing agendas in advance, adhering to written procedures governing the methods used to develop standards, and giving public notice and opportunity for comment on proposed standards.

The ANSI Board of Standards Review verifies that requirements for due process, consensus, and criteria for standards approval are met on a continual basis.

Scope of Committee T1

Committee T1 (also referred to as "the Committee" or "T1") develops standards and technical reports related to interfaces for U.S. telecommunications networks, some of which are associated with other North American telecommunications networks. T1 develops positions on related subjects under consideration in various international standards bodies. T1 focuses on functions and characteristics associated with interconnection and interoperability of telecommunications networks at interfaces with end-user systems, carriers, and information and enhanced-service providers. These include switching, signaling, transmission, performance, operation, administration, and maintenance aspects. Committee T1 is concerned with procedural matters at points of interconnection, such as maintenance and provisioning methods and documentation, for which standardization would benefit the telecommunications industry.

Responsibilities

The T1 Committee is responsible for:

- Developing proposed American national standards;
- Voting on approval of such proposed standards;
- Maintaining and updating the standards developed by the Committee;

- Interpreting the standards developed by the Committee;
- Adopting and revising Committee policies and procedures;
- Other matters requiring Committee action as provided in the Bylaws.

5.2.1.2 T1A1 Performance and Signal Processing

Mission

T1A1 develops and recommends standards and technical reports related to the performance and processing of voice, audio, data, image, and video signals within U.S. telecommunication networks. T1A1 also develops and recommends positions on, and fosters consistency with, standards and related subjects under consideration in other national and international standards bodies.

Scope

The Performance and Signal Processing Technical Subcommittee focuses on two main areas—the performance of networks and services at and between carrier-to-carrier and carrier-to-customer interfaces, with due consideration of end-to-end performance and the performance of customer systems, and signal processing for the transport and integration of voice, audio, data, image, and video signals with due consideration of:

- Interaction with telecommunications networks;
- The integration of inputs and outputs between information processing systems and telecommunication networks;
- Techniques for assessing the performance and impact of such signal processing on telecommunication networks.

Standards, technical reports, and contributions will be developed in accordance with the following objectives.

- To identify and define performance parameters and levels for speed, accuracy, dependability, and availability of connection establishment, information transfer, and connection disengagement (including survivablility parameters)—taking into account characteristics of signal processing systems;
- To define measurement techniques for performance parameters;
- To define methods for characterizing network and signal processing performance for customer applications;

- To identify and develop signal processing algorithms and interface requirements, including coding and compression, interpolation, rate adaptation, echo cancellation, and packetization techniques;
- To take into account interworking of telecommunication networks with customer systems, international networks, other signal processing systems, and new network technologies and services such as asynchronous transfer mode and personal communications.

The T1A1 Committee work is divided into four working groups as defined below.

- T1A1.2: Network Survivability Performance;
- T1A1.3: Digital Network and Service Performance;
- T1A1.5: Multimedia Communications Coding and Performance;
- T1A1.7: Performance and Signal Processing for Voiceband Services.

Multimedia Communications via N-ISDN (H.320)

6

6.1 HISTORY

The ITU is a part of the United Nations, and its purpose is to develop formal "Recommendations" to ensure worldwide communications are accomplished efficiently and effectively. In 1984, the first recommendations for a video teleconferencing codec (H.120 and H.130) were established. These recommendations were defined specifically for the European region (625 lines; 2.048-Mbps primary rate) and for interconnection between Europe and other regions. Since no recommendation existed for nonEuropean regions, it lacked true international scope, and in 1984 the ITU established a "Specialists Group on Coding for Visual Telephony" to develop a truly international recommendation. The ITU established two objectives for the Specialists Group: (1) to develop a recommendation for a video codec for teleconferencing application operating at the bit rates of $N \times 384$ Kbps ($N = 1$ through 5) and (2) to begin the standardization process for a video codec for teleconferencing/video telephone application operating at bit rates of $M \times 64$ Kbps ($M - 1, 2$).

The Specialists Group, chaired by S. Okubo from Japan, met 17 times between 1984 through 1989. At the September meeting in 1988, it was determined that the compression algorithm chosen for $N \times 384$ Kbps was sufficiently flexible so that it could be extended, with good performance, down to 64 Kbps. At that time, the Specialists Group shifted their focus to develop a single recommendation to code at all bit rates from 64 Kbps to 2 Mbps, that is, to code at rates of $p \times 64$ Kbps where the key values of p are 1, 2, 6, 24, and 30.

In 1989 a number of organizations in Europe, the United States, and Japan developed flexible codec systems to meet a preliminary specification of the standard. Various systems were interconnected in the laboratory and by long-distance communication channels to validate the recommendation. These tests were highly successful and encouraging. The ITU formally approved the H.320 series of standards in December 1990.

6.2 H.320: THE VTC SYSTEM STANDARD

H.320 is the overview ITU recommendation that specifies a multimedia terminal for transmission over the N-ISDN network. The entire set of five standards that fully defines the H.320 terminal is listed in Table 6.1.

Since the five H.320 recommendations were finalized by the ITU in December 1990, the standard is extremely mature and forms the cornerstone of the video conferencing business.

Figure 6.1 illustrates the interrelationship of the five H.320 standards and shows the connections of the H.320 terminal with external components. Highlights of key aspects of the H.320 multimedia terminal follow.

One function of the H.320 recommendation is to define the phases of establishing a visual telephone call as listed below.

Table 6.1
H.320 Recommendation Series

Designation	*Title*	*Purpose*
H.320	Narrow-band visual telephone systems and terminal equipment	This standard defines the interrelationship of all of the five H.320 recommendations.
H.261	Video codec for audiovisual services at $P \times 64$ Kbps.	This recommendation specifies the video coding algorithm, the picture format, and forward error correction techniques for the audiovisual terminal.
H.221	Frame structure for a 64-Kbps to 1920-Kbps channel in audio visual teleservices	The purpose of this recommendation is to define a frame structure fo audiovisual teleservices in single or multiple B or HO channels or a single H11 or H12 channel.
H.242	System for establishing communication between audiovisual terminasl using digital channels up to 2 Mbps	Recommendation H.242 defines the detailed "handshake" protocol and procedures that are employed by H.320 terminals in the preliminary phases of a call.
H.230	Frame synchronous *control and indication* (C&I) signals for audiovisual systems	Recommendation H.230 has two primary elements. First, it defines the C& I symbols related to video, audio, maintenance, and multipoint. Second, it contains a table of BAS escape codes that clarifies the circumstances under which some C&I functions are mandatory and others optional.

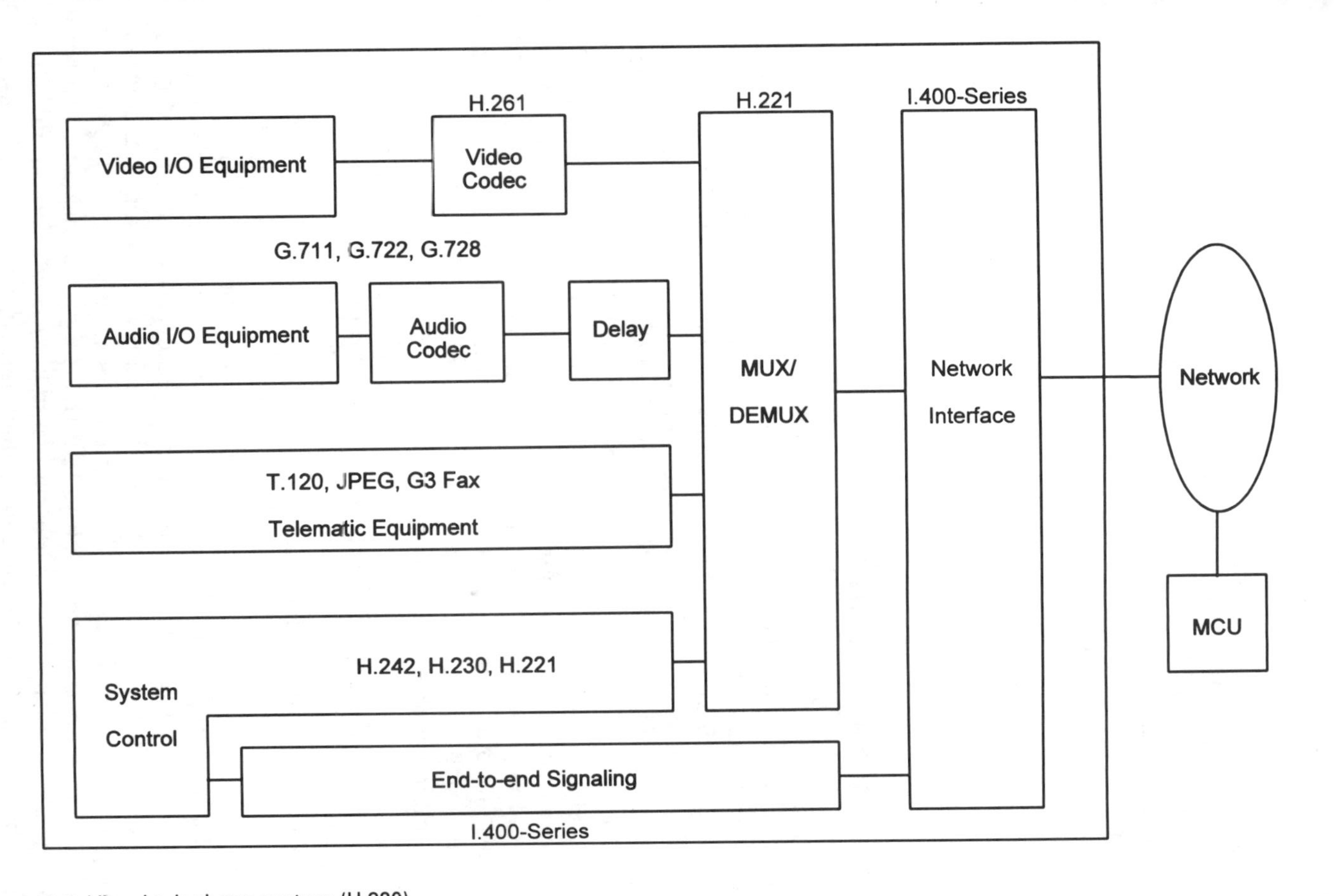

Figure 6.1 Visual telephone system (H.320).

- Phase A: Call set-up, out-band signaling;
- Phase B1: Mode initialization on initial channel;
- Phase CA: Call set-up of additional channel(s), if relevant;
- Phase CB1: Initialization on additional channel(s);
- Phase B2 (or CB2): Establishment of common parameters;
- Phase C: Visual telephone communication;
- Phase D: Termination phase;
- Phase E: Call release.

Another function of Recommendation H.320 is the definition of 16 types of visual telephone terminals and their modes of operation (see Figures 6.2 and 6.3, respectively).

6.3 H.261: VIDEO CODING STANDARD

If the standard TV signal were to be encoded using conventional 8-bit PCM, a bit rate of approximately 90 Mbps would be required for transmission. Video compression technology is used to reduce this bit rate to the primary rates (1.544 Mbps, 2.048 Mbps), fractional primary rates (e.g., 384 Kbps), and basic rates (64 Kbps and multiples), which are employed for economical transmission. The compression function is performed by a video codec (COder, DECoder), and H.261 is the ITU recommendation for the video codec for teleconferencing.

Figure 6.4 is a functional block diagram of the video codec as defined in Recommendation H.261. The heart of the system is the source coder, which compresses the incoming video signal by reducing redundancy inherent in the TV signal. The multiplexer combines the compressed data with various side information, which indicates alternative modes of operation. A transmission buffer is employed to smooth the varying bit rate from the source encoder to adapt it for the fixed bit rate communication channel. A transmission coder includes functions such as forward error control to prepare the signal for the data link.

One of the most challenging problems to be solved by the codec was the reconciliation of the incompatibility between European TV standards (for example, PAL and SECAM) and those in most other areas of the world—NTSC. PAL and SECAM employ 625 lines and a 50-Hz field rate, while NTSC has 525 lines and a 60-Hz field rate. This conflict was resolved by adopting a CIF and QCIF as the picture structure that must be employed for any transmission adhering to H.261. The CIF and QCIF parameters are defined in Table 6.2.

The QCIF format, which employs half the CIF spatial resolution in both horizontal and vertical directions, is the mandatory H.261 format; full CIF is optional. In general, QCIF is used for videophone applications where head-

Visual telephone mode		Channel rate (Kbps)	ISDN channel (NOTE 2)	ISDN interface		Coding	
				Basic	Primary rate	Audio	Video
a	a_0	64	B			Rec. G.711 (NOTE 4)	Rec. H.261 (NOTE 6)
	a_1					Rec. G.728	
b	b_1	128	2B			Rec. G.711	
	b_2					Rec. G.722	
	b_3					Rec. G.728	
q (NOTE 3)	q_1	n x 64	nB			Rec. G.711	
	q_2					Rec. G.722	
	q_3					Rec. G.728	Rec. H.261
g		384	H_0		Applicable	Rec. G.722	
h		768	$2H_0$				
i		1 152	$3H_0$			(NOTE 5)	
j		1 536	$4H_0$	Not applicable			
k		1 536	H_{11}				
l		1 920	$5H_0$				
m		1 920	H_{12}				

NOTE 1 – (Audio coding of mode b_3) In addition to G.728, higher quality audio coding such as H.200/AV.253 may be used for this mode.

NOTE 2 – For multiple channels of B/H_0, all channels are synchronized at the terminal according to 2.7/H.221.

NOTE 3 – q = c/d/e/f corresponds to n = 3/4/5/6, respectively. This mode is applicable to the ISDN basic interface if multiple basic accesses are used.

NOTE 4 – If a visual telephone interworks with a wideband speech terminal, G.722 audio may be used instead of G.711 audio.

NOTE 5 – Modes (G.711 and G.728 audio) other than this recommended mode may be invoked by H.242 procedures.

NOTE 6 – If two terminals connect at this rate and run G.711 and both have video capability, H.261 may be used. It should be noted, however, that the video performance is limited due to the very low bit rate available for this purpose.

Figure 6.2 Communication modes of visual telephone.

and-shoulders pictures are sent from desk to desk. Conversely, the full CIF format is usually used for teleconferencing where several people must be viewed in a conference room.

Figure 6.5 is a functional block diagram outlining the H.261 source coder. Interframe prediction is first carried out in the pixel domain. The prediction errors are encoded by the discrete cosine transform using blocks of 8 pels × 8 pels. The transform coefficients are next quantized and fed to the multiplexer. Motion compensation is included in the prediction on an optional basis.

Mode			Type X (NOTE 2)									Type Y (NOTE 3)					Type Z	
Transfer rate		Audio coding	a	b_1	$b_{2/3}$	b_4	b_5	q_1	$q_{2/3}$	q_4	q_5	1	2	3	4	5		
a_0	B	G.711 (NOTE 4)	X	X	X	X	X	X	X	X	X							
a_1	B	G.728	X	X	X			X	X									
b_1	2B	G.711		X	X	X	X	X	X	X	X							
b_2	2B	G.722			X		X		X		X							
b_3	2B	G.728		X	X			X	X									
q_1	nB	G.711 (NOTE 5)						X	X	X	X							
q_2	nB	G.722 (NOTE 5)							X		X							
q_3	nB	G.728 (NOTE 5)						X	X									
g	H_0	G.722										X	X	X	X	X		
h	$2H_0$	G.722											X	X	X	X		
i	$3H_0$	G.722												X	X	X		
j	$4H_0$	G.722													X	X		
k	H_{11}	G.722															X	
l	$5H_0$	G.722														X		
m	H_{12}	G.722																X

NOTE 1 – "X" means the terminal of the given type is able to work in the given mode.

NOTE 2 – Types Xb_4 and Xb_5 are defined to take into account that G.728 had not yet been recommended when this Recommendation was first established.

NOTE 3 – Terminal of this type must conform to § 3.3.2.2.

NOTE 4 – If a visual telephone interworks with a wideband speech terminal, G.722 audio may be used instead of G.711 audio.

NOTE 5 – q = c/d/e/f corresponds to n = 3/4/5/6, respectively. Since transfer rates of multiple B are defined hierarchically, Type Xf_1, for example, supports all of (a_1, b_1, c_1, d_1, e_1, f_1) and (b_3, c_3, d_3, e_3, f_3) modes.

Figure 6.3 Visual telephone terminal type.

6.3.1 Picture Structure

In the encoding process, each picture is subdivided into *groups of blocks* (GOB). As shown in Figure 6.6, the CIF picture is divided into 12 GOBs while QCIF has only three GOBs. From the GOB level down, the structures of CIF and QCIF are identical. A header at the beginning of the GOB permits resynchronization and changing the coding accuracy.

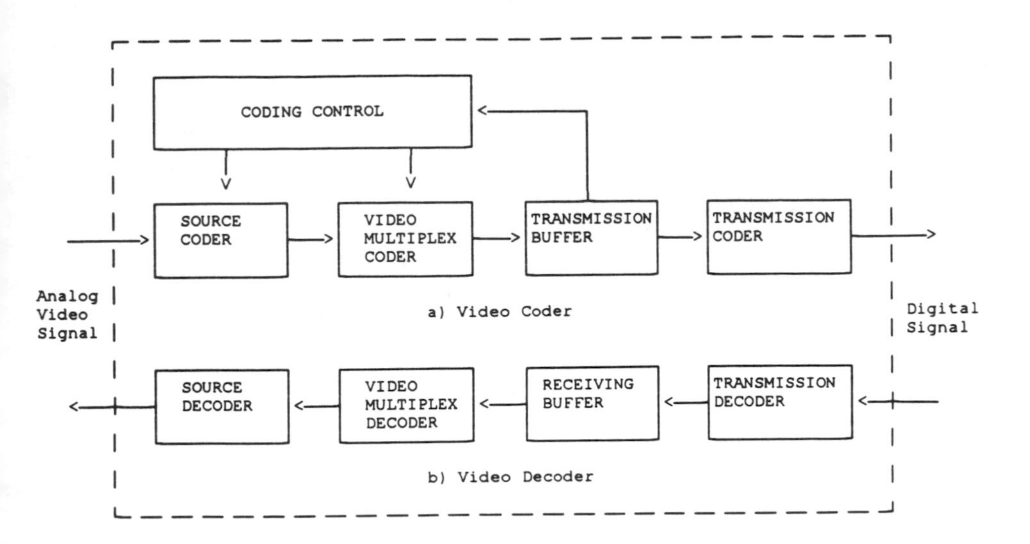

Figure 6.4 Block diagram of the video codec.

Table 6.2
CIF and QCIF Parameters

Parameter	*CIF*	*QCIF*
Coded pictures per second	29.97	(or integral submultiples)
Coded luminance pixels per line	352	176
Coded luminance lines per picture	288	144
Coded color pixels per line	176	88
Coded color lines per picture	144	72

Each GOB is further divided into 33 macroblocks, as shown in Figure 6.7. The macroblock header defines the location of the macroblock within the GOB, the type of coding to be performed, possible motion vectors, and which blocks within the macroblock will actually be coded. There are two basic types of coding. In Intra coding, coding is performed without reference to previous pictures. This mode is relatively rare but is required for forced updating, and every macroblock must occasionally be Intra coded to control the accumulation of inverse transform mismatch errors. The more common coding type is Inter, in which only the difference between the previous and the current pictures is coded. Of course, for picture areas without motion, the macroblock does not have to be coded at all.

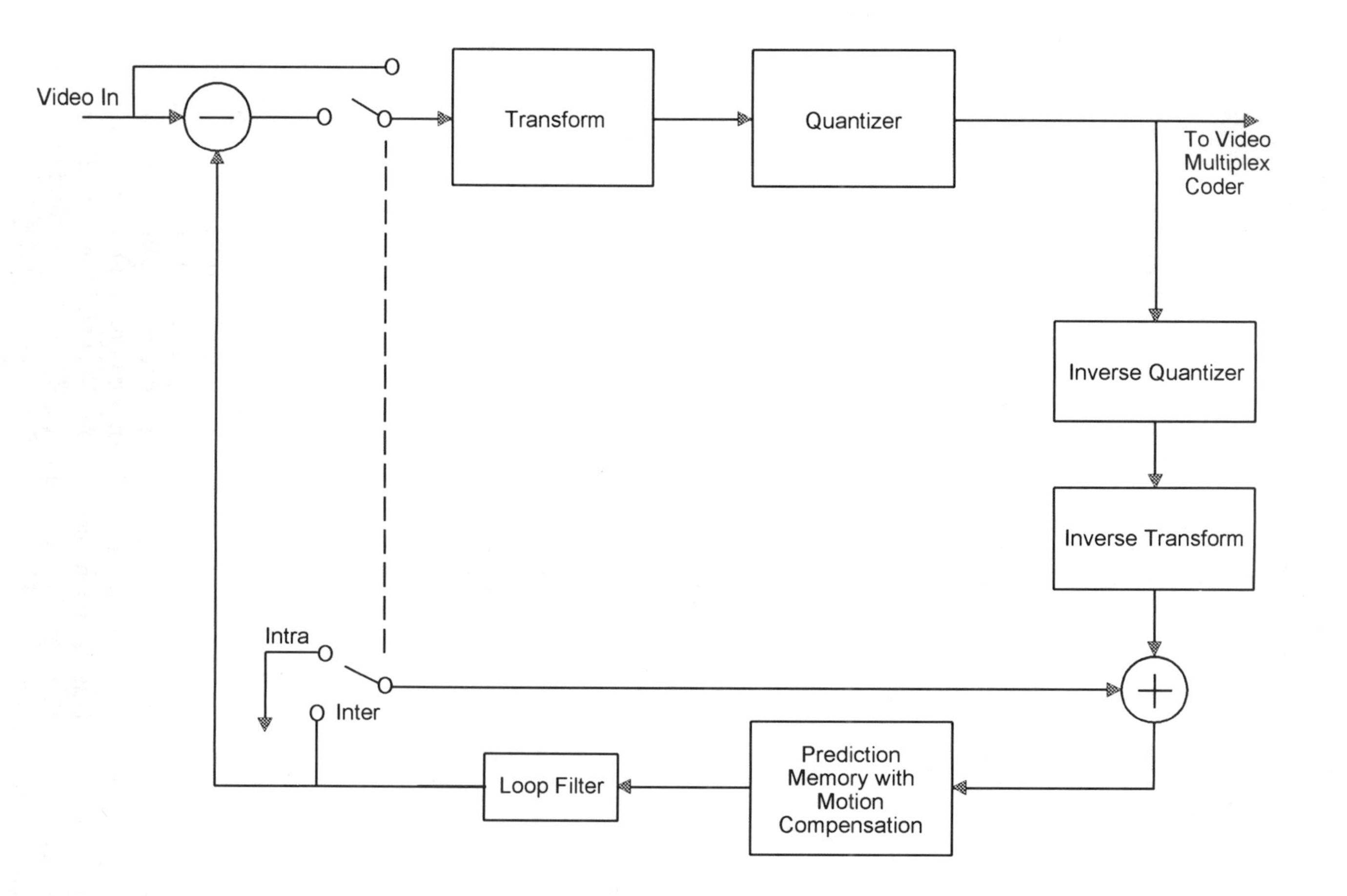

Figure 6.5 Source coder.

1	2
3	4
5	6
7	8
9	10
11	12

CIF

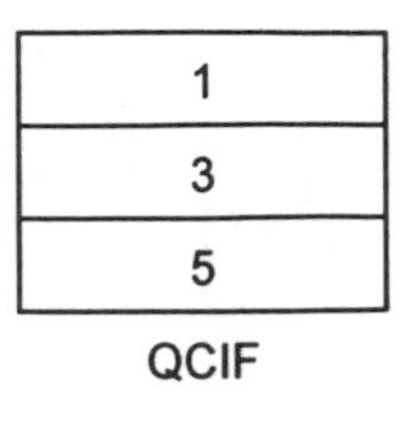

Figure 6.6 Arrangement of GOBs in a picture.

1	2	3	4	5	6	7	8	9	10	11
12	13	14	15	16	17	18	19	20	21	22
23	24	25	26	27	28	29	30	31	32	33

Figure 6.7 Arrangement of macroblocks in a GOB.

Each macroblock is further divided into six blocks, as shown in Figure 6.8. Four of the blocks represent the luminance, or brightness, while the other two represent the red and blue color differences. Each block is 8 by 8 pixels, so it can be seen that the color resolution is half of the luminance resolution in both dimensions.

6.3.2 Example of Block Coding

Figure 6.9 shows a simple example of how each 8 × 8 block is coded. In this case, Intra coding is used, but the principle is the same for Inter coding. Figure 6.9(a) shows the original block to be coded. Without compression, this would take 8 bits to code each of the 64 pixels, or a total of 512 bits. First, the

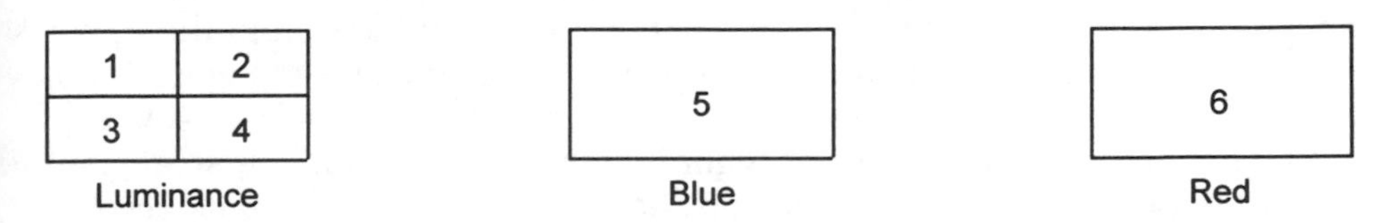

Figure 6.8 Arrangement of blocks in a macroblock.

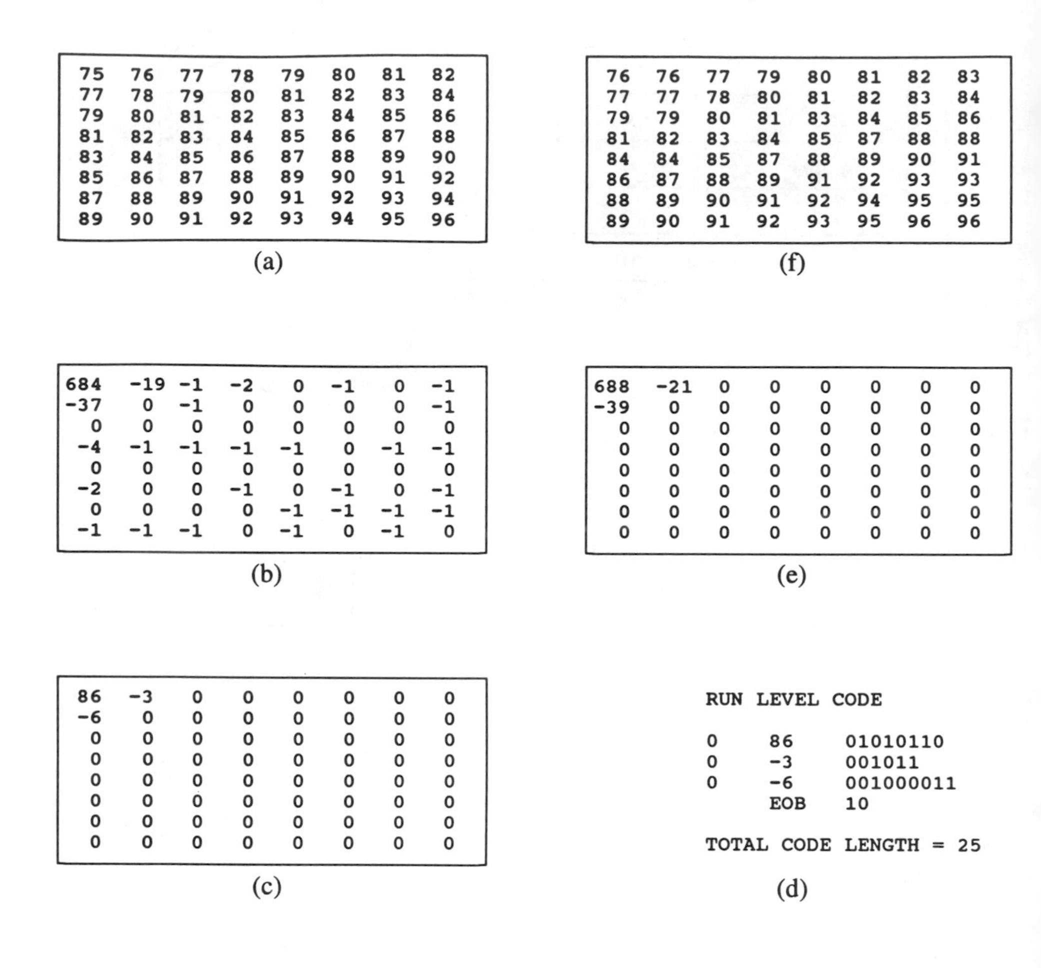

75	76	77	78	79	80	81	82
77	78	79	80	81	82	83	84
79	80	81	82	83	84	85	86
81	82	83	84	85	86	87	88
83	84	85	86	87	88	89	90
85	86	87	88	89	90	91	92
87	88	89	90	91	92	93	94
89	90	91	92	93	94	95	96

(a)

76	76	77	79	80	81	82	83
77	77	78	80	81	82	83	84
79	79	80	81	83	84	85	86
81	82	83	84	85	87	88	88
84	84	85	87	88	89	90	91
86	87	88	89	91	92	93	93
88	89	90	91	92	94	95	95
89	90	91	92	93	95	96	96

(f)

684	-19	-1	-2	0	-1	0	-1
-37	0	-1	0	0	0	0	-1
0	0	0	0	0	0	0	0
-4	-1	-1	-1	-1	0	-1	-1
0	0	0	0	0	0	0	0
-2	0	0	-1	0	-1	0	-1
0	0	0	0	-1	-1	-1	-1
-1	-1	-1	0	-1	0	-1	0

(b)

688	-21	0	0	0	0	0	0
-39	0	0	0	0	0	0	0
0	0	0	0	0	0	0	0
0	0	0	0	0	0	0	0
0	0	0	0	0	0	0	0
0	0	0	0	0	0	0	0
0	0	0	0	0	0	0	0
0	0	0	0	0	0	0	0

(e)

86	-3	0	0	0	0	0	0
-6	0	0	0	0	0	0	0
0	0	0	0	0	0	0	0
0	0	0	0	0	0	0	0
0	0	0	0	0	0	0	0
0	0	0	0	0	0	0	0
0	0	0	0	0	0	0	0
0	0	0	0	0	0	0	0

(c)

RUN	LEVEL	CODE
0	86	01010110
0	-3	001011
0	-6	001000011
	EOB	10

TOTAL CODE LENGTH = 25

(d)

Figure 6.9 Sample intra block coding: (a) original block (8 × 8 × 8 = 512 bits), (b) transformed block coefficients, (c) quantized coefficient levels, (d) coefficients in zig-zag order and variable length code, (e) inverse quantized coefficients, and (f) reconstituted block.

block is transformed, using the two-dimensional DCT, giving the coefficients of Figure 6.9(b). Note that most of the energy is concentrated into the upper left-hand corner of the coefficient matrix. Next, the coefficients of Figure 6.9(b) are quantized with a step size of 6. (The first term, DC, always uses a step size of 8.) This produces the values of Figure 6.9(c), which are much smaller in magnitude than the original coefficients, and most of the coefficients become zero. The larger the step size, the smaller the values produced, resulting in more compression.

The coefficients are then reordered, using the zigzag scanning order of Figure 6.10. All zero coefficients are replaced with a count of the number of zero's before each nonzero coefficient (RUN). Each combination of RUN and VALUE produces a *variable length code* (VLC) that is sent to the decoder. The last nonzero VALUE is followed by an *end of block* (EOB) code. The total number of bits used to describe the block is 25, a compression of 20:1.

At the decoder (and at the coder to produce the prediction picture), the step size and VALUEs are used to reconstruct the inverse quantized coefficients, which, as shown in Figure 6.9(e) are similar to, but not exactly equal to, the original coefficients. When these coefficients are inverse transformed, the result of Figure 6.9(f) is obtained. Note that the differences between this block and the original block are quite small.

6.3.3 Motion Compensation

The operation of motion compensation is shown in Figure 6.11. Block "A" is a block in the current picture that is to be coded. Block "B" is the block at the same position as "A" but in the picture that was previously stored in both coder and decoder. Because of image motion, block "A" more closely resembles the pixel data from block "C" than that from block "B". The displacement of block "C" from block "B", measured in pixels in the x- and y-directions, is the motion vector. The pixel-by-pixel difference between blocks "A" and "C" is transformed and coded. The motion vector and code data are transmitted to the decoder, where the inverse transformed block data is added to the data in block "C" pointed to by the motion vector and placed in the block "A" position.

1	2	6	7	15	16	28	29
3	5	8	14	17	27	30	43
4	9	13	18	26	31	42	44
10	12	19	25	32	41	45	54
11	20	24	33	40	46	53	55
21	23	34	39	47	52	56	61
22	35	38	48	51	57	60	62
36	37	49	50	58	59	63	64

Figure 6.10 Scanning order in a block.

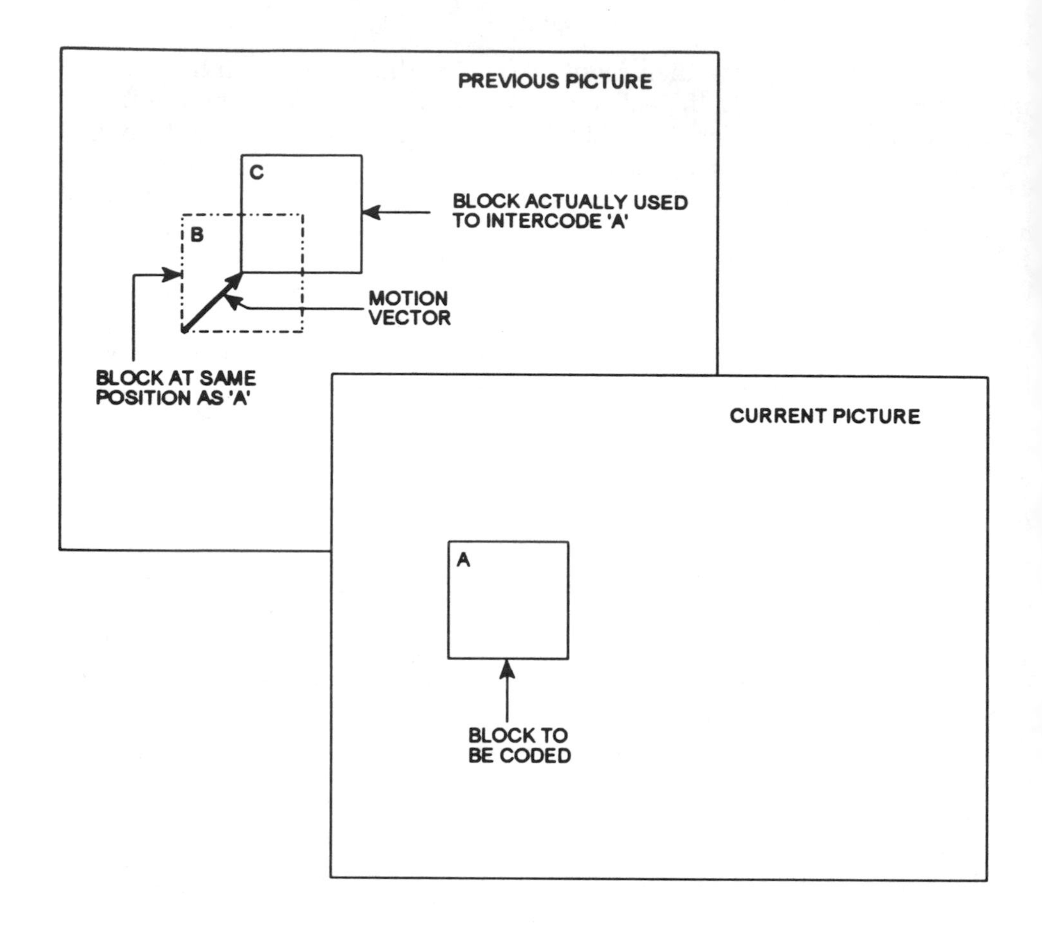

Figure 6.11 Interframe coding with motion vectors.

The use of motion vectors is optional in the coder, where the calculation of the optimum motion vectors is complex, but required in the decoder, where the reconstruction of the motion is relatively simple.

The H.261 standard does not define all aspects of image coding and decoding. Rather it is just an interoperability specification, guaranteeing that any codec manufactured according to the standard will be able to communicate with each other. This still allows considerable freedom for manufacturers to offer better performance, and new developments may be able to be incorporated. (This is in contrast with the G.722 audio standard, where the coding algorithm is rather precisely defined.) For example, the encoder strategy is not defined. Which blocks will be encoded, with what type of code, and with what accuracy is under control of the designer. While there is less freedom for the decoder, post processing, such as filtering or interpolation of the image, is under control of the designer.

Furthermore, the H.242 Recommendation permits two codecs to negotiate to an altogether different proprietary algorithm that they both incorporate, with the H.261 algorithm becoming a fall-back mode for codecs of different manufacture.

6.4 H.221: FRAME STRUCTURE FOR A 64-Kbps TO 20-Kbps CHANNEL IN AUDIOVISUAL TELESERVICES

The purpose of this recommendation is to define a frame structure for audiovisual teleservices in single or multiple B or HO channels or a single H11 or H12 channel that makes the best use of the characteristics and properties of the audio and video encoding algorithms, of the transmission frame structure, and of the existing ITU recommendations. It offers several advantages, as follows.

- It is simple, economic, and flexible. It may be implemented on a simple microprocessor using well-known hardware principles.
- It is a synchronous procedure. The exact time of a configuration change is the same in the transmitter and the receiver. Configurations can be changed at 20 ms intervals.
- It needs no return link for audiovisual signal transmission, since a configuration is signaled by repeatedly transmitted codewords.
- It is very secure in case of transmission errors, since the code controlling the multiplex is protected by a double-error correcting code.
- It allows the synchronization of multiple 64-Kbps or 384-Kbps connections and the control of the multiplexing of audio, video, data, and other signals within the synchronized multiconnection structure in the case of multimedia services such as videoconference.

This recommendation provides for dynamically subdividing an overall transmission channel of 64 Kpbs to 1920 Kbps into lower rates suitable for audio, video, data, and telematic purposes. The overall transmission channel is derived by synchronizing and ordering transmissions over from 1B to 6B connections, from 1HO to 5HO connections, or an H11 or H12 connection.

A single 64-Kbps channel is structured into octets transmitted at 8 kHz. Each bit position of the octets may be regarded as a subchannel of 8 Kbps. The eighth subchannel is called the *Service Channel* (SC), containing the two critical parts.

- *Frame Alignment Signal* (FAS): This 8-bit code is used to frame the 80 octets of information in a B channel.

- *Bit-Rate Allocation Signal* (BAS): This 8-bit code describes the capability of a terminal to structure the capacity of the channel or synchronized multiple channels in various ways and to command a receiver to demultiplex and make use of the constituent signals in such structures. This signal is also used for controls and indications.

The video bitstream is carried in frames of data as shown in Figure 6.12. Each frame corresponds to a 64-Kbps B channel in ISDN. Two frames are shown—one for the audio portion of the conference, and the other for the video portion. In each, there is an 8-bit FAS that permits synchronization of the frame and low-speed signaling of communication overhead. There is also an 8-bit BAS that defines how the H.221 channels and subchannels are divided and what type of service is used on each section. For example, one BAS code is used for "Standard Video to Rec. H.261," while another might indicate that two B channels are allocated to this service. The BAS codes can change from frame to frame to indicate complex protocols or changes of mode.

Each frame of 640 bits is transmitted in 10 ms, giving an overall rate of 64,000 bits per second. However, the FAS and BAS use 16 of the 640 bits, so the net rate available for video is only 62.4 Kbps for a single B channel. The order of transmission is left to right across each row and then the row below. Higher bit rates can be obtained by using multiple B channels (up to 2 for ISDN basic access, up to 30 for primary access).

6.5 H.242: SYSTEM FOR ESTABLISHING COMMUNICATION BETWEEN AUDIOVISUAL TERMINALS USING DIGITAL CHANNELS UP TO 2 MEGABITS PER SECOND

Recommendation H.242 defines the detailed *handshake* protocol and procedures that are employed by H.320 terminals in the preliminary phases of a call. Major topics covered in this recommendation are:

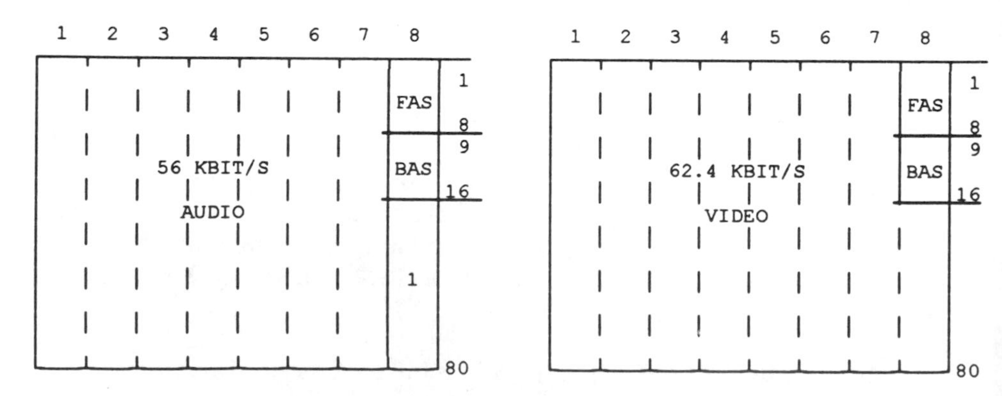

Figure 6.12 H.221 frames.

- Basic sequences for in-channel procedures;
- Mode initialization, dynamic mode switching, and mode O forcing recovery from fault conditions;
- Network consideration: call connections, disconnection, and call transfer;
- Procedures for activation and deactivation of data channels;
- Procedures for operation of terminals in restricted networks.

6.6 H.230: FRAME SYNCHRONOUS CONTROL AND INDICATION SIGNALS FOR AUDIOVISUAL SYSTEMS

Digital audiovisual services are provided by a transmission system in which the relevant signals are multiplexed onto a digital path. In addition to the audio, video, user data, and telematic information, these signals include information for the proper functioning of the system. The additional information has been named *control and indication* (C&I) to reflect the fact that while some bits are genuinely for *control*, causing a state change somewhere else in the system, others provide for indications to the users as to the functioning of the system.

Recommendation H.230 has two primary elements. First, it defines the C&I symbols related to video, audio, maintenance, and multipoint. Second, it contains a table of BAS escape codes that clarifies the circumstances under which some C&I functions are mandatory and others optional.

6.7 AUDIO CODING

The BAS codes of H.221 are used to signal a wide range of possible audio coding modes. The most prominent modes define existing ITU Recommendations G.711, G.722, and G.728 (see Table 6.3). Recommendation G.711 (Pulse Code Modulation of Voice Frequencies) is used for narrowband speech since it samples only at 8,000 samples/s and encodes to 8 bits/sample for a transmission rate of 64 Kbps.

Table 6.3
ITV Audio Coding Standards

CCITT Recommendation	*Audio Bandwidth (kHz)*	*Bit Rate (Kbps)*	*Coding Algorithm*
G.711	3	64	PCM
G.722	7	64	Sub-band adaptive
		56	differential PCM
		48	
G.728	3	16	CELP

Recommendation G.722 (7-kHz audio coding with 64 Kbps) describes the characteristics of an audio (50 Hz to 7,000 Hz) coding system that may be used for a variety of higher quality speech applications, The coding system uses *sub-band adaptive differential pulse code modulation* (SB-ADPCM) within a bit rate of 64 Kbps. In the SB-ADPCM technique used, the frequency band is split into two sub-bands (higher and lower) and the signals in each sub-band are encoded using ADPCM. The system has three basic modes of operation corresponding to the used bit rates of 7-kHz audio coding: 64 Kbps, 56 Kbps, and 48 Kbps.

The G.7ll and G.722 standards have been used extensively in $P \times 64$ systems operating at high bit rates (for example, 1.544 Mbps and 768 Kbps) where the 64 Kbps for audio is a relatively small percentage of the transmission bit rate. However, in recent years, the channel bit rate has been typically reduced to 112 Kbps, where 64 Kbps audio would require too large a fraction of the channel capacity. The recent G.728 standard operating at 16 Kbps is a major development that alleviates this crowding.

6.8 DATA CHANNEL

The H.320 standards provide for the transmission of data along with video and audio information. *Low-speed data* (LSD) and *high-speed data* (HSD) channels are specified. See Figure 6.13. These operate up to 64 Kbps and at multiples of 64 Kbps, respectively. For these channels to be useful it is necessary to develop communication protocols and applications to define data flow. Figure 6.13 shows that the T.126 (still image/annotation) and T.127 (binary file transfer) application protocols have been developed for use over the T.120 communication protocol stack and the *Multilayer Protocol* (MLP) data channel as defined by the ITU Study Group 8. The H.280 (Far End Camera Control) application protocol has also been developed for use over the H.224 (Data Link Layer) communications protocol.

6.9 MULTIPOINT

In 1992, the two ITU recommendations that were approved were H.231, Multipoint Control Unit for Audiovisual Systems Using Digital Channels up

APPLICATIONS		T.121 - Audiograph Conferencing T.126 - Still Image T.res - Reservations T.avc - Audio & Video Control T.bwc - Bandwidth Control T.tdc - Transparent Data Channels T.fax - Multipoint Fax T.pro - Terminal Profiles T.127 - Binary File Transfer	H.280 - Far End Camera Control (Simplex, low delay)
COMMUNICATION PROTOCOL		T.120 Series - T.123 Protocol Stack - T.122 Multipoint Comm. Svc. - T.124 Conference Control - T.125 Multipoint Protocol	H.224 - Data Link Layer (DLL)
DATA CHANNELS	Low Speed (<64 Kbps)	MLP (4, 6.4, VAR)	LSD (Low Speed Data)
	High Speed (Multiples of 64 Kbps)	H-MLP	HSD (High Speed Data)

Figure 6.13 Data transmission in H.320.

to 2 Mbps, and H.243, System for Establishing Communication Between Three or More Audiovisual Terminals Using Digital Channels up to 2 Mbps.

As the name implies, H.231 defines a multipoint control unit (MCU) that serves as a bridge in multipoint connections. Recommendation H.243 defines the communication protocol between an H.320 ($P \times 64$) terminal and an H.231 MCU. The MCU enables three or more H.320 terminals to participate in an audiovisual conference. Two or more MCUs can be cascaded as illustrated in Figure 6.14. The MCU provides audio mixing and a video switching capability that provides for video teleconferencing.

In multipoint systems, the user may be presented with continuous presence (sometimes called "Hollywood squares") or switched displays. Continuous presence involves spatially multiplexing the selected images into a single image in a split screen format. In the switched mode a user views the signal originating from one of the other conference participants. There are three primary ways to control the switching process to determine which of the other users is viewed: voice activated switching, user selection, chair control. In the case of voice activation, the speaker's image is automatically presented to all

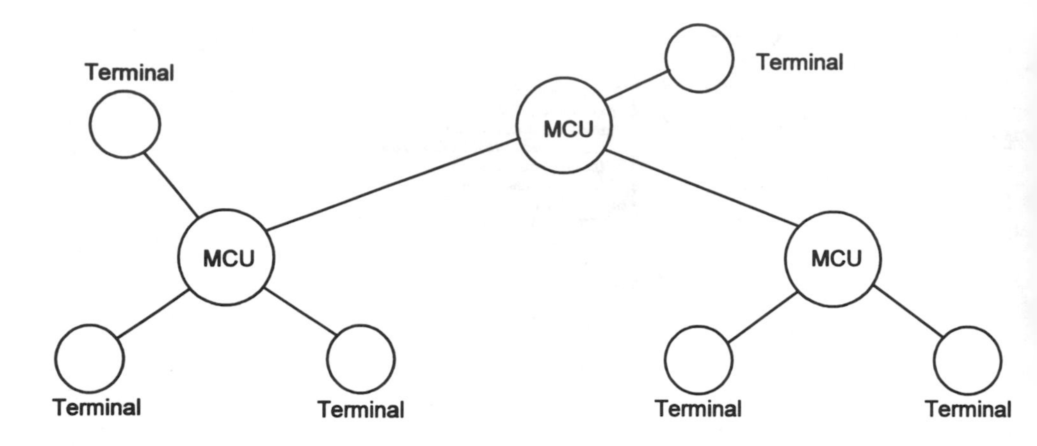

Figure 6.14 Typical multipoint configuration.

conferees. In the case of chair control, the conference chairman has control over which participant is viewed by all users.

6.10 PRIVACY

The ITU has developed two recommendations to provide for the privacy of transmission between H.320 audiovisual terminals. A privacy system consists of the confidentiality mechanism or encryption process for the data and a key management subsystem.

H.233 describes the confidentiality part of a privacy system suitable for H.320 videoconferencing terminals. Although an encryption algorithm is required for such a privacy system, the specification of such an algorithm is not included. The recommendations refer to a number of possible algorithms (for example, DES, FEAL, and BCRYPT), and more are being added. H.233 will be completed in two phases. Phase I for point-to-point encryption was completed in 1993, while multipoint encryption will be defined later.

Recommendation H.234 describes the authentication and key management methods for a privacy system suitable for H.230 terminals. Privacy is achieved by using secret keys. The keys are loaded into the confidentiality part of the privacy system and control the way in which the transmitted data is encrypted and decrypted. If a third party gains access to the keys being used, then the privacy system is no longer secure.

6.11 NARROWBAND ISDN (N-ISDN)

The digital pipe between the central office and the ISDN subscriber will be used to carry a number of communication channels. The capacity of the pipe, and therefore the number of channels carried, may vary from user to user. The transmission structure of any access link will be constructed from the following types of channels:

- B channel: 64 Kbps;
- D channel: 16 Kbps or 64 Kbps;
- H channel: 384 Kbps, 1536 Kbps, or 1920 Kbps.

The B channel is a user channel that can be used to carry digital data, PCM-encoded digital voice, or a mixture of lower rate traffic including digital data and digitized voice encoded at a fraction of 64 Kbps. In the case of mixed traffic, all traffic of the B channel must be destined for the same endpoint; that is, the elemental unit of circuit switching is the B channel.

The D channel serves two main purposes. First, it carries common-channel signaling information to control circuit-switched calls on associated B channels at the user interface. In addition, the D channel may be used for packet-switching or low-speed (for example, 100 bps) telemetry at times when no signaling information is waiting.

H channels are provided for user information at higher bit rates. The user may use such a channel as a high-speed trunk or subdivide the channel according to the user's own TDM scheme. Examples of applications include fast facsimile, video, high-speed data, high-quality audio, and multiplexed information streams at lower data rates.

These channel types are grouped into transmission structures that are offered as a package to the user. The best-defined structures at this time are the basic channel structure basic access and the primary channel structure primary access.

6.11.1 Basic Access

Basic access consists of two full-duplex 64-Kbps B channels and a full-duplex 16-Kbps D channel. The total bit rate, by simple arithmetic, is 144 Kbps. However, framing, synchronization, and other overhead bits bring the total bit rate on a basic access link to 192 Kbps. The basic service is intended to meet the needs of most individual users, including residential subscribers and very small offices. It allows the simultaneous use of voice and other applications, such as video conferencing. Most existing two-wire local loops can support this interface.

6.11.2 Primary Access

Primary access is intended for users with greater capacity requirements, such as offices with a digital PBX or a LAN. Because of differences in the digital transmission hierarchies used in different countries, it was not possible to get agreement on a single data rate. The United States, Canada, and Japan make use of a transmission structure based on 1.544 Mbps; this corresponds to the T1 transmission facility of AT&T. In Europe, 2.048 Mbps is the standard rate. Both of these data rates are provided as a primary interface service. Typically, the channel structure for the 1.544-Mbps rate will be 23 B channels plus one 64-Kbps D channel and for the 2.048-Mbps rate, 30 B channels plus one 64-Kbps D channel. Again, it is possible for a customer with lesser requirements to employ fewer B channels, in which case the channel structure is nB + D, where n ranges from 1 to 23 or from 1 to 30 for the two primary services. Also, a customer with high data rate demands may be provided with more than one primary physical interface. In this case, a single D channel on one of the interfaces may suffice for all signaling needs and the other interfaces may consist solely of B channels (24B or 31B).

The primary interface may also be used to support H channels. Some of these structures include a 64-Kbps D channel for control signaling. When no D channel is present, it is assumed that a D channel on another primary interface at the same subscriber location will provide any required signaling. The following structures are recognized.

- *Primary rate interface H0 channel structures:* This interface supports multiple 384-Kbps H0 channels. The structures are 3H0 + D and 4H0 for the 1.544-Mbps interface and 5H0 +D for the 2.048-Mbps interface.
- *Primary rate interface H1 channel structures:* The H11 channel structure consists of one 1536-Kbps H11 channel. The H12 channel structure consists of one 1920-Kbps H12 channel and one D channel.
- *Primary rate interface structures for mixtures of B and H0 channels:* Consists of zero or one D channels plus any possible combination of B and H0 channels up to the capacity of the physical interface (for example, 3H0 + 5B + D and 3H0 + 6B).

Multimedia Communications via the PSTN and Mobile Radio (H.324)

7

7.1 OVERVIEW

In September 1993 the ITU established a program to develop an international standard for a videophone terminal operating over the *public switched telephone network* (PSTN). A major milestone in this project was accomplished in March 1996, when the ITU approved the standard. It is anticipated that the H.324 terminal will have two principal applications, namely, a conventional videophone used primarily by the consumer and a multimedia system to be integrated into a personal computer for a range of business purposes (for example, telecommuting).

In addition to approving the umbrella H.324 Recommendation, the ITU has completed the four major functional elements of the terminal: the G.723.1 speech coder, the H.263 video coder, the H.245 communication controller, and the H.223 multiplexer. The quality of the speech provided by the new G.723.1 audio coder, when operating at only 6.4 Kbps, is very close to that found on a conventional phone call. The picture quality produced by the new H.263 video coder shows promise of significant improvement relative to many earlier systems. It is anticipated that these technical advances, when combined with the high transmission bit rate of the V.34 modem (28.8 Kbps maximum), will yield an overall audiovisual system performance that is significantly improved relative to earlier videophone terminals.

At the same meeting in Geneva, the ITU announced the acceleration of the schedule to develop a standard for a videophone terminal to operate over mobile radio networks. The new terminal, designated AV.324/M, will be based upon the design of the H.324 device to ease interoperation between the mobile and telephone networks.

7.2 H.324: TERMINAL FOR LOW-BIT RATE MULTIMEDIA COMMUNICATION

Recommendation H.324 describes terminals for low-bitrate multimedia communication, utilizing V.34 modems operating over the GSTN. H.324 terminals may carry real-time voice, data, and video, or any combination, including videotelephony.

H.324 terminals may be integrated into personal computers or implemented in stand-alone devices such as videotelephones. Support for each media type (such as voice, data, and video) is optional, but if supported, the ability to use a specified common mode of operation is required so that all terminals supporting that media type can interwork. H.324 allows more than one channel of each type to be in use. Other recommendations in the H.324 series include the H.223 multiplex, H.245 control, H.263 video codec, and G.723.1 audio codec.

H.324 makes use of the logical channel signaling procedures of Recommendation H.245, in which the content of each logical channel is described when the channel is opened. Procedures are provided for expression of receiver and transmitter capabilities, so that transmissions are limited to what receivers can decode and that receivers may request a particular desired mode from transmitters. Since the procedures of H.245 are also planned for use by Recommendation H.310 for ATM networks and Recommendation H.323 for nonguaranteed bandwidth LANs, interworking with these systems should be straightforward.

H.324 terminals may be used in multipoint configurations through MCUs and may interwork with H.320 terminals on the ISDN as well as with terminals on wireless networks. H.324 implementations are not required to have each functional element, except for the V.34 modem, H.223 multiplex, and H.245 system control protocol, which shall be supported by all H.324 terminals.

H.324 terminals offering audio communication shall support the G.723.1 audio codec. H.324 terminals offering video communication shall support the H.263 and H.261 video codecs. H.324 terminals offering real-time audiographic conferencing should support the T.120 protocol suite. In addition, other video and audio codecs and other data protocols may optionally be used via negotiation over the H.245 control channel.

If a modem external to the H.324 terminal is used, terminal/modem control shall be according to V.25ter.

Multimedia information streams are classified into video, audio, data, and control as follows.

- Video streams are continuous traffic-carrying moving color pictures. When used, the bitrate available for video streams may vary according to the needs of the audio and data channels.
- Audio streams are real-time but may optionally be delayed in the receiver processing path to maintain synchronization with the video streams. To reduce the average bitrate of audio streams, voice activation may be provided.

- Data streams may represent still pictures, facsimile, documents, computer files, computer application data, undefined user data, and other data streams.
- Control streams pass control commands and indications between remotelike functional elements. Terminal-to-modem control is according to V.25ter for terminals using external modems connected by a separate physical interface. Terminal-to-terminal control is according to H.245.

The H.324 document refers to other ITU recommendations, as illustrated in Figure 7.1, that collectively define the complete terminal. Four new companion recommendations include H.263 (Video Coding for Low Bitrate Communication), G.723.1 (Speech Coder for Multimedia Telecommunications Transmitting at 5.3/6.3 Kbps), H.223 (Multiplexing Protocol for Low-Bitrate Multimedia Terminals), and H.245 (Control of Communications Between Multimedia

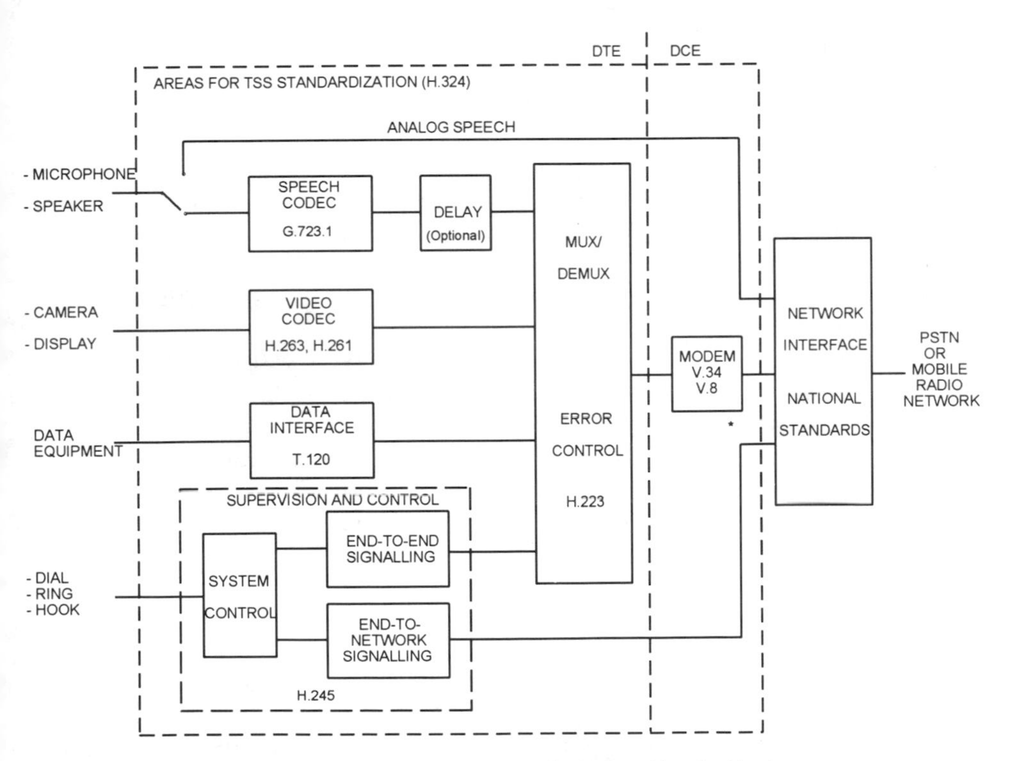

Figure 7.1 Functional block diagram for very low bitrate videophone.

Terminals). H.324 specifies use of the V.34 modem, which operates up to 28.8 Kbps, and the V.8 (or V.8bis) procedure to start and stop data transmission. An optional data channel is defined to provide for the exchange of computer data in the workstation/PC environment. H.324 specifies the use of the T.120 protocol as one possible means for this data exchange.

Recommendation H.324 defines the seven phases of a call: set-up, speech only, modem training, initialization, message, end, and clearing.

7.3 G.723.1: SPEECH CODER FOR MULTIMEDIA TELECOMMUNICATIONS TRANSMITTING AT 5.3/6.3 Kbps

All H.324 terminals offering audio communication shall support both the high and low rates of the G.723.1 audio codec. G.723.1 receivers shall be capable of accepting silence frames. The choice of which rate to use is made by the transmitter and is signaled to the receiver in-band in the audio channel as part of the syntax of each audio frame. Transmitters may switch G.723.1 rates on a frame-by-frame basis, based on bitrate, audio quality, or other preferences. Receivers may signal, via H.245, a preference for a particular audio rate or mode.

Alternative audio codecs may also be used, via H.245 negotiation. Coders may omit sending audio signals during silent periods after sending a single frame of silence or may send silence background fill frames if such techniques are specified by the audio codec recommendation in use.

More than one audio channel may be transmitted, as negotiated via the H.245 control channel.

The G.723.1 speech coder can be used for a wide range of audio signals but is optimized to code speech. The system's two mandatory bit rates are 5.3 Kbps and 6.3 Kbps. The coder is based on the general structure of the *Multipulse-Maximum Likelihood Quantizer* (MP-MLQ) speech coder. The MP-MLQ excitation will be used for the high-rate version of the coder. The *Algebraic Codebook Excitation Linear Prediction* (ACELP) excitation is used for the low-rate version. The coder provides a quality essentially equivalent to that of a POTS toll call. For clear speech, or with background speech, the 6.3-Kbps mode provides speech quality equivalent to the 32-Kbps G.726 coder. The 5.3-Kbps mode performs better than the IS54 digital cellular standard.

Performance of the coder has been demonstrated by extensive subjective testing. The speech quality in reference to 32-Kbps G.726 ADPCM (considered equivalent to toll quality) and 8-Kbps IS54 VSELP is given in Table 7.1. This table is based on a subjective test conducted for the French language. In all cases the performance of G.726 was rated better than or equal to IS54. All tests were conducted with four talkers except the speaker variability test, for which

Table 7.1
Results of Subjective Test for G.723.1

	High Rate	*Low Rate*
Speaker variability	= G.726	= IS54
One encoding	= G.726	= IS54
Tandem	> 4*G.726	> 4*G.726
Level +10 dB	= G.726	= G.726
Level –10 dB	> G.726	> G.726
Frame erasures (3%)	= G.726 – 0.5	= G.726 – 0.75
Flat input (loudspeaker)	= G.726	= G.726
Flat in (loudspeaker) 2T	> 4*G.726	> 4*G.726
Office noise (18 dB) DMOS	= IS54	= IS54
Babble noise (20 dB) DMOS	= G.726	> IS54
Music noise (20 dB) DMOS	> IS54	< IS54

twelve talkers were used. The symbols <, =, and > are used to identify less than, equivalent to, and better than, respectively. Comparisons were made by taking into account the statistical error of the test. The background noise conditions are speech signals mixed with the specified background noise.

From these results one can conclude that within the scope of the test, both low- and high-rate coders are always equivalent to or better than IS54, except for the low-rate coder with music, and that the high rate coder is always equivalent to G.726, except for office and music background noises.

The complexity of the dual rate coder depends on the DSP chip and the implementation but is approximately 18 Mips and 16 Mips for 6.3 Kbps and 5.3 Kbps, respectively. The memory requirements for the dual rate coder are

RAM: 2240 16-bit words;
ROM: 9100 16-bit words (tables),
7000 16-bit words (program).

The algorithmic delay is 30-ms frame + 7.5-ms look ahead, resulting in 37.5 ms.

G.723.1 can be integrated with any voice activity detector to be used for speech interpolation or discontinuous transmission schemes. Any possible extensions would need agreements on the proper procedures for encoding low-level background noises and comfort noise generation.

7.4 H.263: VIDEO CODING FOR LOW-BIT RATE COMMUNICATION

All H.324 terminals offering video communication shall support both the H.263 and H.261 video codecs, except H.320 Interworking Adapters, which are not terminals and do not have to support H.263 (see Section 7.2). The H.261 and H.263 codecs shall be used without BCH error correction and without error correction framing. The five standardized image formats are 16CIF, 4CIF, CIF, QCIF, and SQCIF.

CIF and QCIF are defined in H.261. For the H.263 algorithm, SQCIF, 4CIF, and 16CIF are defined in H.263. For the H.261 algorithm, SQCIF is any active picture size less than QCIF, filled out by a black border, and coded in the QCIF format. For all these formats, the pixel aspect ratio is the same as that of the CIF format. Table 7.2 shows which picture formats are required and which are optional for H.324 terminals that support video.

All video decoders shall be capable of processing video bit streams of the maximum bitrate that can be received by the implementation of the H.223 multiplex (for example, maximum V.34 rate for single link and 2 × V.34 rate for double link).

Which picture formats, minimum number of skipped pictures, and algorithm options can be accepted by the decoder are determined during the capability exchange using H.245. After that, the encoder is free to transmit anything that is in line with the decoder's capability. Decoders that indicate capability for a particular algorithm option shall also be capable of accepting video bit streams that do not make use of that option.

The H.263 coding algorithm is an extension of H.261. H.263 describes, as H.261 does, a hybrid DPCM/DCT video coding method. Both standards use techniques such as DCT, motion compensation, variable length coding, and

Table 7.2
Picture Formats for Video Terminals

Picture Format	*Luminance Pixels*	*Encoder*		*Decoder*	
		H.261	H.263	H.261	H.263
SQCIF	128 X 96 for H.263‡	Optional‡	Required*,†	Optional‡	Required*
QCIF	176 X 144	Required	Required*,†	Required	Required*
CIF	352 X 288	Optional	Optional	Optional	Optional
4CIF	704 X 576	Not defined	Optional	Not defined	Optional
16CIF	1408 X 1152	Not defined	Optional	Not defined	Optional

*Optional for H.320 interworking adapters.
†It is mandatory to encode one of the picture formats QCIF and SQCIF; it is optional to encode both formats.
‡H.261 SQCIF is any active size less than QCIF, filled out by a black border and coded in QCIF format.

scalar quantization and both use the well-known macroblock structure. Differences between H.263 and H.261 are:

- PB frames (optional);
- Overlapped block motion compensation (optional);
- Motion vectors pointing outside the picture (optional);
- 8 pel × 8 pel motion vectors (optional);
- Syntax-based arithmetic coding (optional);
- H.263 has an optional GOB level;
- H.263 uses different VLC tables at the Macroblock and block levels;
- H.263 uses half pel motion compensation instead of full pel plus loop filter;
- In H.263, there is no still picture mode (JPEG is used for still pictures);
- In H.263, there is no error detection/correction included like the BCH in H.261;
- H.263 uses a different form of macroblock addressing;
- H.263 does not use the end of block marker.

Of particular interest is the optional PB-frame mode. A PB-frame consists of two pictures being coded as one unit. The name PB comes from the name of picture types in MPEG where there are P-pictures and B-pictures. Thus a PB-frame consists of one P-picture that is predicted from the last decoded P-picture and one B-picture that is predicted both from the last decoded P-picture and the P-picture currently being decoded. This last picture is called a B-picture because parts of it may be bidirectionally predicted from the past and future P-pictures. The prediction process is illustrated in Figure 7.2.

It has been shown that the H.263 system typically outperforms H.261 (when adapted for the GSTN application) by 2.5 to 1. This means that when adjusted to provide equal picture quality, the H.261 bitrate is approximately 2.5 times that for the H.263 codec.

7.5 H.245: CONTROL PROTOCOL FOR MULTIMEDIA COMMUNICATIONS

The control channel carries end-to-end control messages governing the operation of the H.324 system, including capabilities exchange, opening and closing of logical channels, mode preference requests, multiplex table entry transmission, flow control messages, and general commands and indications.

There shall be exactly one control channel in each direction within H.324, which shall use the messages and procedures of Recommendation H.245. The control channel shall be carried on logical channel 0. The control channel shall be considered to be permanently open from the establishment of digital

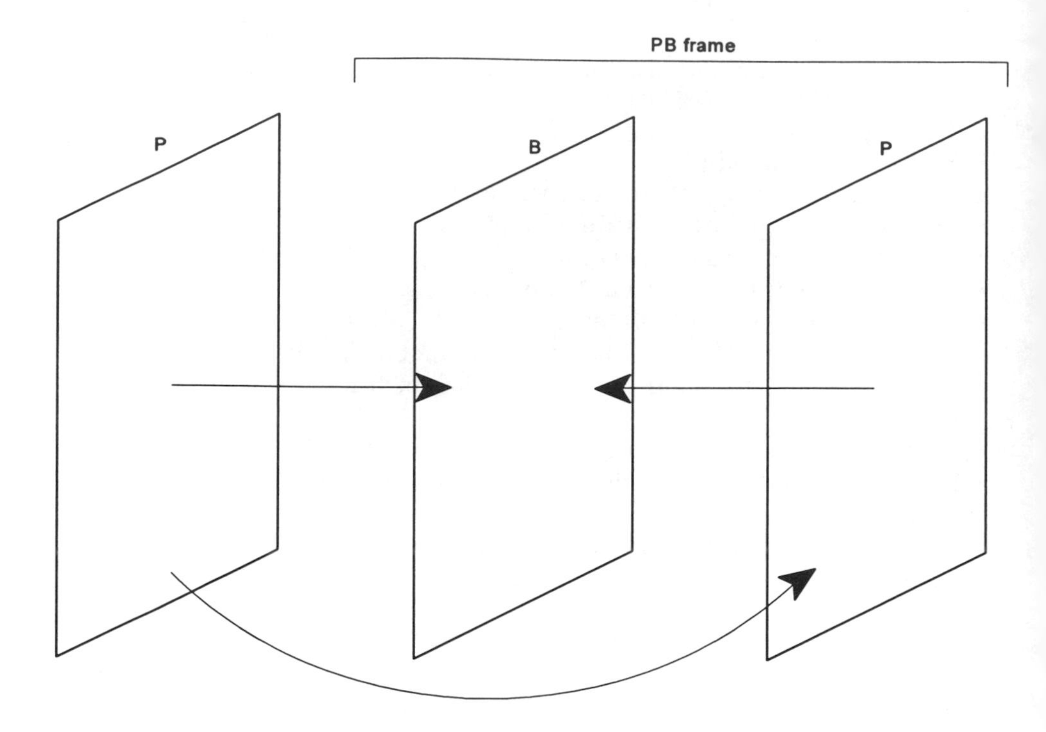

Figure 7.2 Prediction in PB-frames mode.

communication until the termination of digital communication; the normal procedures for opening and closing logical channels shall not apply to the control channel.

General commands and indications shall be chosen from the message set contained in H.245. In addition, other command and indication signals may be sent that have been specifically defined to be transferred in-band within video, audio, or data streams (see the appropriate recommendation to determine if such signals have been defined).

H.245 messages fall into four categories—Request, Response, Command, and Indication. Request messages require a specific action by the receiver, including an immediate response. Response messages respond to a corresponding request. Command messages require a specific action but do not require a response. Indication messages are informative only and do not require any action or response. H.324 terminals shall respond to all H.245 commands and requests as specified in H.245 and shall transmit accurate indications reflecting the state of the terminal.

Table 7.3 shows how the total bit rate available from the modem might be divided into its various constituent virtual channels by the H.245 control

Table 7.3
Example of a Bitrate Budget for Very Low Bitrate Visual Telephony (Kilobits per Second)

		Virtual Channel			
	Modem Bit Rate	*Overhead/ Supervision (5%)*	*Speech*	*Video*	*Data*
	9.6 Kbps	0.5 Kbps	4.8 Kbps	4.3 Kbps	Variable
			6.8	2.3	
Overall transmission bit rate	14.4	0.7	4.8	8.9	Variable
			6.8	6.9	
	[†] ⋮	⋮	⋮	⋮	
	21.6	1.1	4.8	16.7	Variable
			6.8	14.7	
	[†] ⋮	⋮	⋮	⋮	
	28.8	1.4	4.8	22.6	Variable
			6.8	20.6	
Virtual channel bit rate characteristic		Variable bit rate	Dedicated, fixed bit rate[*]	Variable bit rate	Variable bit rate
Priority[‡]		High priority	High priority	Lowest priority	Higher than video, lower than overhead/ speech

[*]The plan includes consideration of advanced speech codec technology such as a dual bit rate speech codec and a reduced bit rate when voiced speech is not present.
[†]V.34 operates at increments of 2.4 Kbps, that is, 16.8, 19.2, 21.6, 24.0, 26.4, 28.8 Kbps.
[‡]The channel priorities will not be standardized; the priorities indicated are examples.

system. The overall bit rates are those specified in the V.34 modem. Note that V.34 can operate at increments of 2.4 Kbps up to 28.8 Kbps; only 21.6 Kbps and 28.8 Kbps are shown in the table as examples. Overhead/Supervision is shown as a hypothetical value of 5% of the overall bit rate. This includes the Supervisory channel and packet headers and error control bits. Speech is shown for two bit rates that are representative of possible speech coding rates. The video bit rate shown is what is left after deducting the Overhead/Supervision and Speech bit rates from the Overall Transmission bit rate. The Data would take a variable number of bits from the Video, either a small amount or all the

Video bits, depending on the designer's or the user's control. Provision is made for both point-to-point and multipoint operation. Recommendation H.245 creates a flexible, extensible infrastructure for a wide range of multimedia applications including storage/retrieval, messaging, and distribution services as well as the fundamental conversational use. The control structure is applicable to the situation where only data and speech are transmitted (without motion video) as well as the case where speech, video, and data are required.

7.6 H.223: MULTIPLEXING PROTOCOL FOR LOW-BIT RATE MULTIMEDIA COMMUNICATION

This recommendation specifies a packet-oriented multiplexing protocol designed for the exchange of one or more information streams between higher layer entities such as data and control protocols and audio and video codecs that use this recommendation.

In this recommendation, each information stream is represented by a unidirectional logical channel that is identified by a unique *logical channel number* (LCN). LCN 0 is a permanent logical channel assigned to the H.245 control channel. All other logical channels are dynamically opened and closed by the transmitter using the H.245 OpenLogicalChannel and CloseLogicalChannel messages. All necessary attributes of the logical channel are specified in the OpenLogicalChannel message. For applications that require a reverse channel, a procedure for opening bidirectional logical channels is also defined in H.245. The general structure of the multiplexer is shown in Figure 7.3. The multiplexer consists of two distinct layers, particularly, a *multiplex* (MUX) layer and an *adaptation layer* (AL).

7.6.1 Multiplex Layer

The MUX layer is responsible for transferring information received from the AL to the far end using the services of an underlying physical layer. The MUX layer exchanges information with the AL in logical units called MUX-SDUs, which always contain an integral number of octets that belong to a single logical channel. MUX-SDUs typically represent information blocks whose start and end mark the location of fields that need to be interpreted in the receiver.

MUX-SDUs are transferred by the MUX layer to the far end in one or more variable length packets called MUX-PDUs. MUX-PDUs consist of the HDLC opening flag, followed by a one-octet header and followed by a variable number of octets in the information field that continue until the closing HDLC flag (see Figures 7.4 and 7.5). The HDLC zero-bit insertion method is used to ensure that a flag is not simulated within the MUX-PDU.

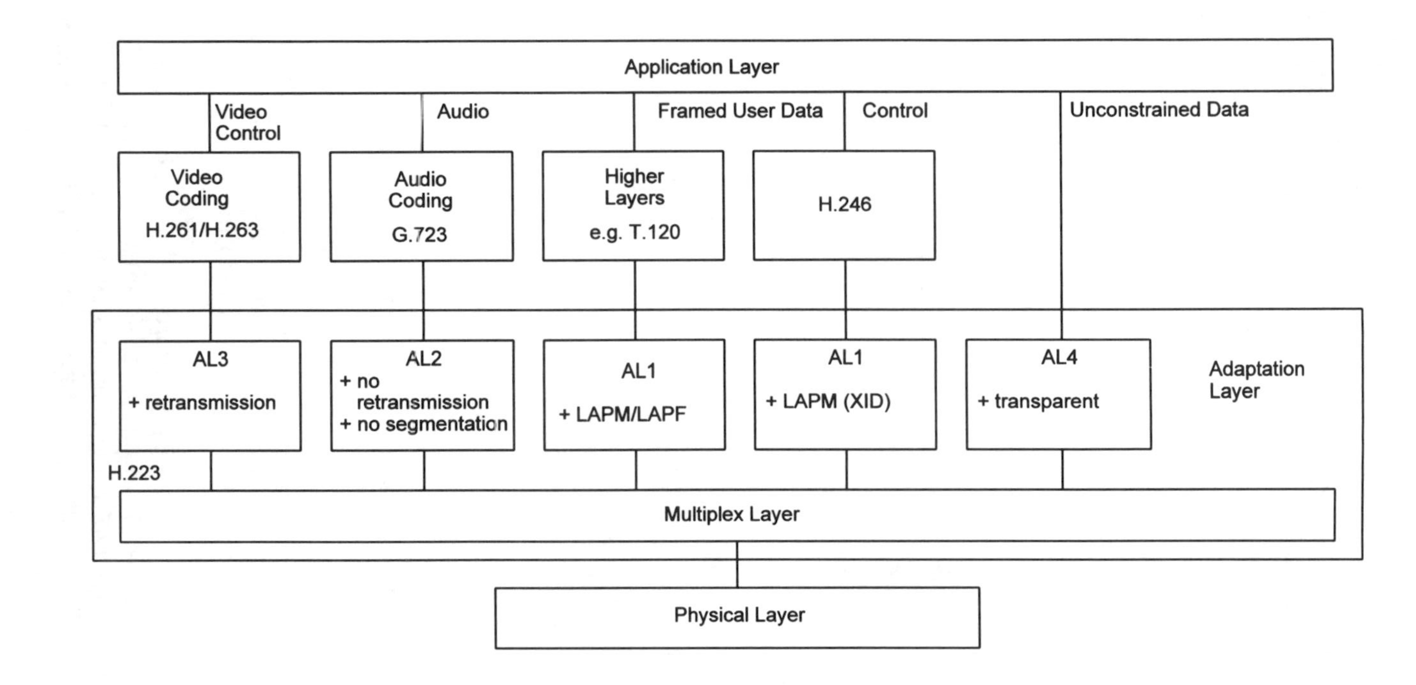

Figure 7.3 Protocol structure of H.223.

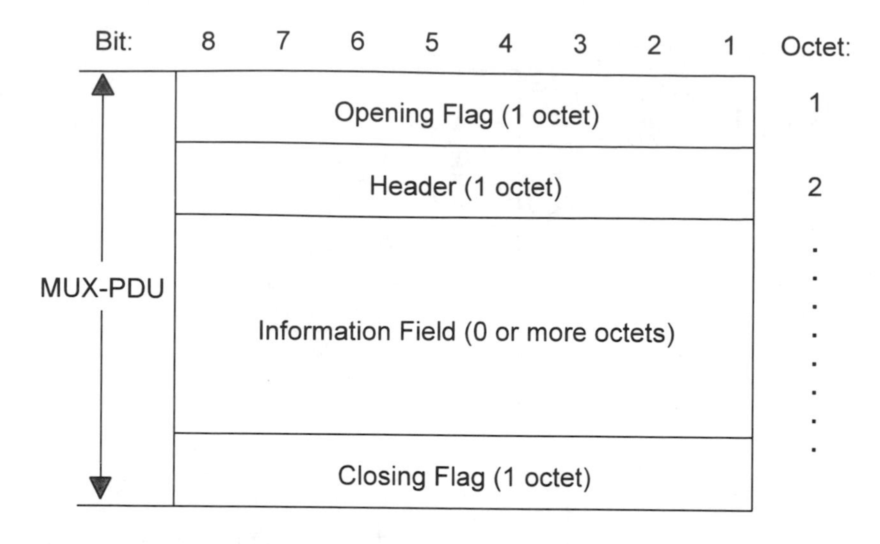

Figure 7.4 MUX-PDU format.

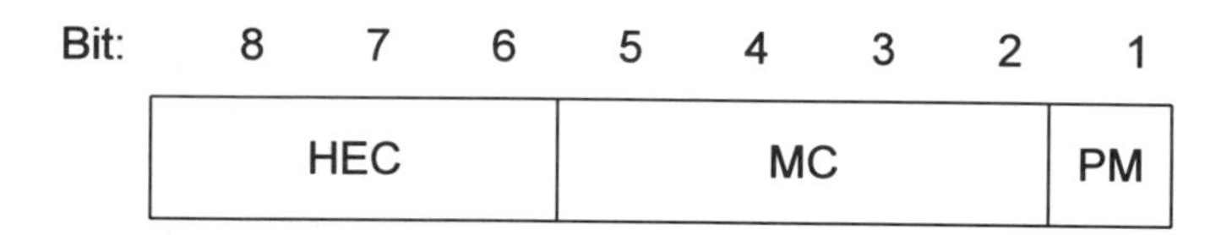

MC: Multiplex Code
HEC: Header Error Control
PM: Packet Marker

Figure 7.5 Header format of the MUX-PDU.

Octets from multiple logical channels may be present in a single MUX-PDU information field. The header octet contains a 4-bit *multiplexes code* (MC) field that specifies, by reference to a multiplex table entry, the logical channel to which each octet in the information field belongs. Multiplex table entry 0 is permanently assigned to the control channel. Other multiplex table entries are formed by the transmitter and are signaled to the far end via the control channel prior to their use.

Multiplex table entries specify a pattern of slots each assigned to a single logical channel. Any one of 16 multiplex table entries may be used in any given MUX-PDU. This allows rapid low-overhead switching of the number of bits allocated to each logical channel from one MUX-PDU to the next. The construction of multiplex table entries and their use in MUX-PDUs is entirely under the control of the transmitter, subject to certain receiver capabilities.

7.6.2 Adaptation Layer

The unit of information exchanged between the AL and the higher layer AL users is an AL-SDU. The method of mapping information streams from higher layers into AL-SDUs is outside the scope of this recommendation and is specified in the System Recommendation that uses H.223.

AL-SDUs contain an integer number of octets. The AL adapts AL-SDUs to the MUX layer by adding, where appropriate, additional octets for purposes such as error detection, sequence numbering, and retransmission. The logical information unit exchanged between peer AL entities is called an AL-PDU. An AL-PDU carries exactly the same information as a MUX-SDU.

Three different types of ALs, named AL1 through AL3, are specified in this recommendation.

AL1 is designed primarily for the transfer of data or control information. Since AL1 does not provide any error control, all necessary error protection should be provided by the AL1 user. In the framed transfer mode, AL1 receives variable length frames from its higher layer (for example, a data link layer protocol such as LAPM/V.42 or LAPF/Q.922, which provides error control) in AL-SDUs and simply passes these to the MUX layer in MUX-SDUs without any modifications. In the unframed mode, AL1 is used to transfer an unframed sequence of octets from an AL1 user. In this mode, one AL-SDU represents the entire sequence and is assumed to continue indefinitely.

AL2 is designed primarily for the transfer of digital audio. AL2 receives frames, possibly of variable length, from its higher layer (for example, an audio encoder) in AL-SDUs and passes these to the MUX layer in MUX-SDUs, after adding one octet for an 8-bit CRC and optionally adding one octet for sequence numbering.

AL3 is designed primarily for the transfer of digital video. AL3 receives variable length frames from its higher layer (for example, a video encoder) in AL-SDUs and passes these to the MUX layer in MUX-SDUs, after adding two octets for a 16-bit CRC and optionally adding one or two control octets. AL3 includes a retransmission protocol designed for video.

An example of how audio, video, and data fields could be multiplexed by the H.223 systems is illustrated in Figure 7.6.

7.7 DATA CHANNEL

All data channels are optional. Standardized options for data applications include:

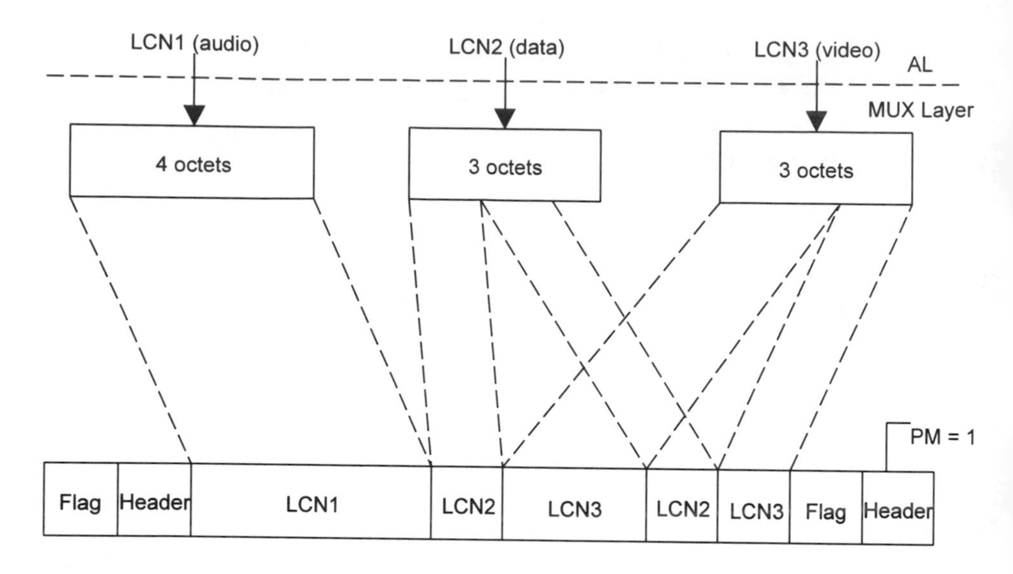

Figure 7.6 Information field example.

- T.120 series for point-to-point and multipoint audiographic teleconferencing including database access, still image transfer and annotation, application sharing, and real-time file transfer;
- T.84 (SPIFF) point-to-point still image transfer cutting across application borders;
- T.434 point-to-point telematic file transfer cutting across application borders;
- H.224 for real-time control of simplex applications, including H.281 far-end camera control;
- Network link layer, per ISO/IEC TR9577 (supports IP and PPP network layers, among others);
- Unspecified user data from external data ports.

These data applications may reside in an external computer or other dedicated device attached to the H.324 terminal through a V.24 or equivalent interface (implementation dependent) or may be integrated into the H.324 terminal itself. Each data application makes use of an underlying data protocol for link layer transport. For each data application supported by the H.324 terminal, this recommendation requires support for a particular underlying data protocol to ensure interworking of data applications.

The H.245 control channel is not considered a data channel. Standardized link layer data protocols used by data applications include:

- Buffered V.14 mode for transfer of asynchronous characters, without error control;
- LAPM/V.42 for error-corrected transfer of asynchronous characters (additionally, depending on application, V.42bis data compression may be used);
- HDLC frame tunneling for transfer of HDLC frames;
- Transparent data mode for direct access by unframed or self-framed protocols.

All H.324 terminals offering real-time audiographic conferencing should support the T.120 protocol suite.

7.8 AUDIOVISUAL TELEPHONY VIA MOBILE RADIO

In February 1995 the ITU requested that the LBC Experts Group begin work to adapt the H.324 series of GSTN recommendations for application to mobile networks. Figure 7.7 illustrates the general structure of the H.324M mobile multimedia terminal. Work toward the H.324M Recommendation has been divided into the following areas of study: (1) speech error protection, (2) video error protection, (3) communications control (adjustments to H.245), (4) multiplex/error control of the multiplexed signal, and (5) system.

Several of the general principles and underlying assumptions upon which the H.324M Recommendations are to be based are as follows.

- H.324M Recommendations should be based upon H.324 as much as possible.
- The technical requirements and objectives for H.324M are essentially the same as for H.324.
- Since the vast majority of mobile terminal calls are with terminals in fixed networks, it is very important that H.32M Recommendations be developed to maximize interoperability with these fixed terminals.
- It is assumed that the H.324M terminal has access to a transparent/ synchronous bit stream from the mobile network.
- It is proposed to provide the manufacturer of mobile multimedia terminals with a number of optional error protection tools to address a wide range of mobile networks, that is, regional and global, present and future, and cordless and cellular. Consequently H.324M tools should be flexible, bitrate scaleable, and extensible to the maximum degree possible.
- Like H.324, nonconversational services are an important application for H.324M.

Considerable progress has been made toward video error protection for H.324M. Technology that appears particularly promising includes ARQ, GOB

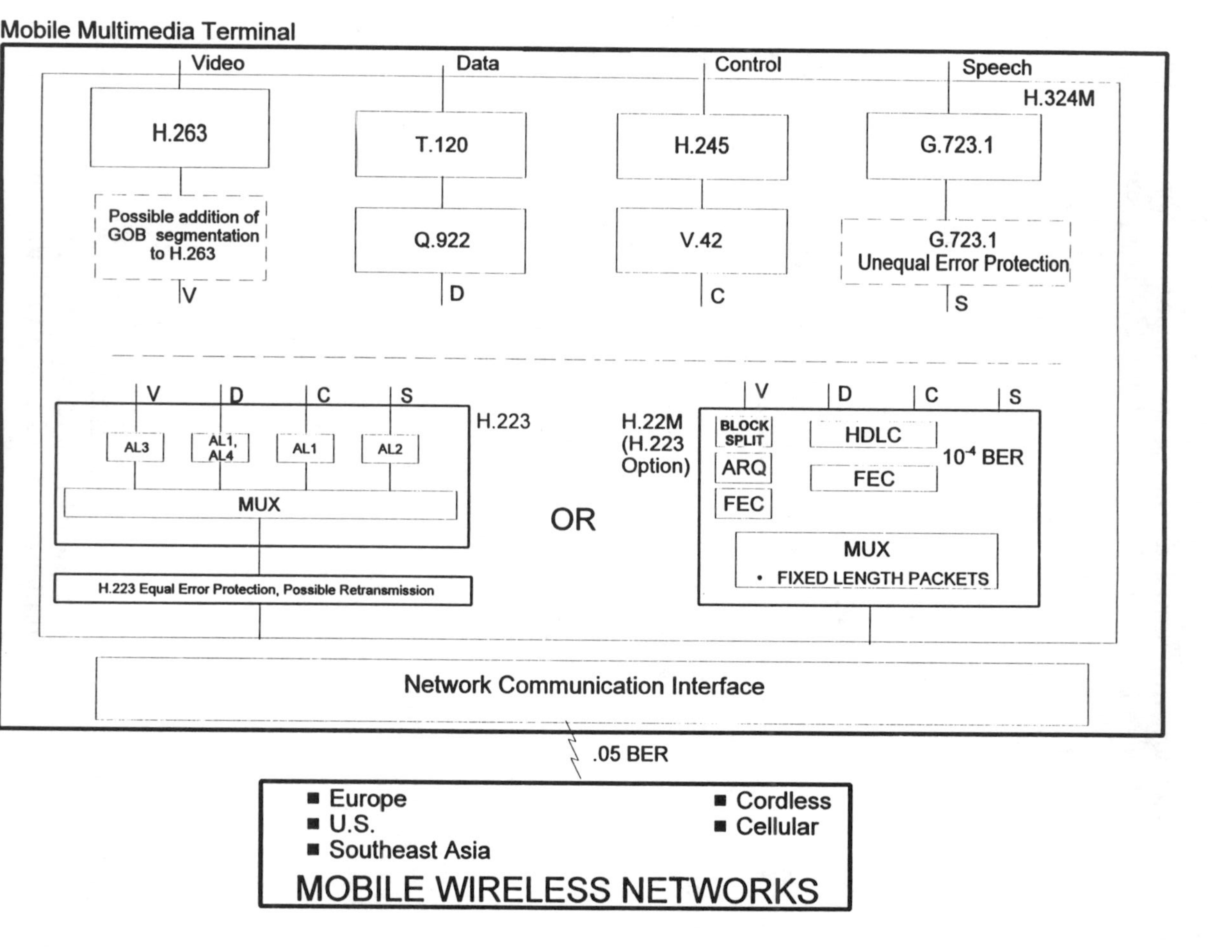

Figure 7.7 Mobile multimedia terminal.

segmentation, and error concealment. Two different approaches for the H.324M multiplexer are being planned. The first, which will be mandatory, merely applies error protection to the H.223 multiplexed signal. The second, which will be optional, employs a completely new multiplex structure designed specifically for transmission over mobile networks with severe error conditions. Work is also underway to determine the value of scaleable FEC/Interleaving for protection of video errors. A scalable channel coder has been defined to adapt the G.723.1 speech coder to mobile channels. It will be proposed for decision in May 1996.

Multimedia Transmission via B-ISDN (H.321, H.310) and LAN Networks (H.322, H.323)

8

8.1 BROADBAND ISDN

ISDNs are divided into narrowband and broadband parts. N-ISDN operates at rates equal to or less than the primary rates (for example, 1.544 Mbps), while the B-ISDN operates at rates above the primary rates.

Broadband aspects of the ISDN (B-ISDN) are being developed by ITU-T Study Group 18 to establish a customer-switched digital network. SG18 decided to standardize the *network node interface* (NNI) by a worldwide unique *synchronous digital hierarchy* (SDH). This was achieved by Working Party 7, which is responsible for transmission aspects of digital networks. Figure 8.1 illustrates the worldwide unique NNI. The SDH specifies 155.52 Mbps as the worldwide unique interface bit rate. The proposal of Study Group 18 for B-ISDN as described in Recommendation I.121 is that the target transfer mode is the *asynchronous transfer mode* (ATM) in which the data is transmitted in a series of fixed-size blocks called cells (see Figure 8.2). Packet-switched networks already exist for the transmission of digital data for non-real-time services (for example, the exchange of information between computer databases). In this instance, if a packet is corrupted or lost, the receiving terminal can request that the particular packet be retransmitted. Recommendation I.121, however, envisages that the B-ISDN will carry all the telecommunications services provided in the future including real-time services such as telephony, videoconferencing, and videophony, as well as television and sound contribution and distribution services. For these real-time services, if a cell is corrupted or lost, retransmission of cells is not possible and so degradation of the signal may occur.

The main advantage claimed for ATM is that the network switches are no longer bit rate and service specific; in the B-ISDN all services (including future new, and as yet unspecified services) are expected to be carried, and a common user-network interface will exist for all services. Many of the important

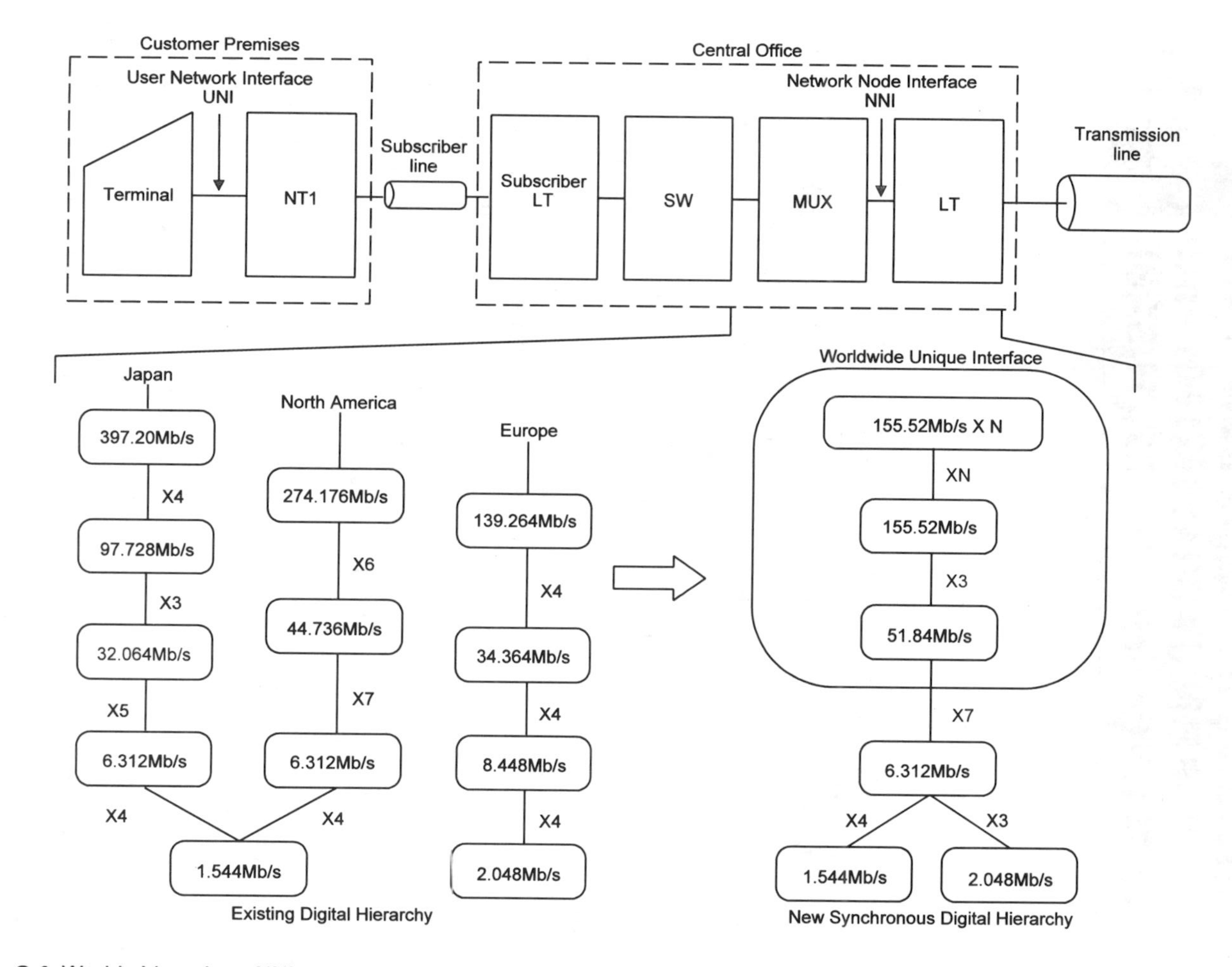

Figure 8.1 Worldwide unique NNI.

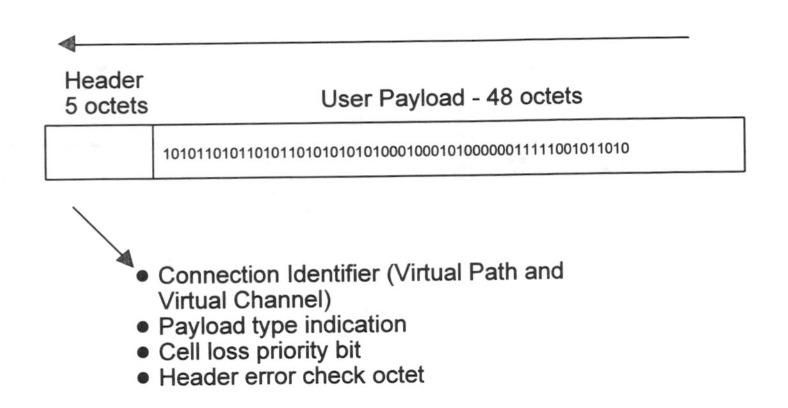

Figure 8.2 ATM cell.

parameters of B-ISDN have still to be specified. However, an ATM-based network will introduce some effects not experienced in synchronous networks, such as cell delay jitter and occasional cell loss.

An ATM-based network will, in principle, provide the user with whatever bit rate is required (within the constraints of the interface and the network) so that teleconference users, for example, could decide on the optimum picture quality required by sessions. Additionally, new television services at different bit rates could be transmitted over the network through the same user-network interface. With continuing improvement in picture coding algorithms and with advances in technology allowing more complex algorithms to be implemented, service providers could, in the future, offer either an improved quality of service at the same average bit rate or the same quality of service at a lower average bit rate. An ATM-based network will in principle have the flexibility to provide additional transmission capacity when required and could allow the development of a variable bit rate/constant quality coding scheme.

An ATM connection consists of a concatenation of ATM layer links. The ATM layer does no processing of the user payload. Figure 8.3 shows an example of an ATM connection in terms of the B-ISDN Protocol Reference Model. Service specific functions reside above the solid line at the network edge.

At ATM virtual channel connection can be viewed as a cell pipe. Cells arrive at the receiver in the order in which they were transmitted but with some delay variation. ATM errors are such that some cells may be missing due to congestion in the network and some cells may be incorrectly sent to the receiver due to bit errors in the ATM cell header. In addition, the user data in some cells may contain errors.

ATM operates in a connection-oriented mode. At connection set-up, the user of the ATM layer negotiates a traffic contract with the network. The network

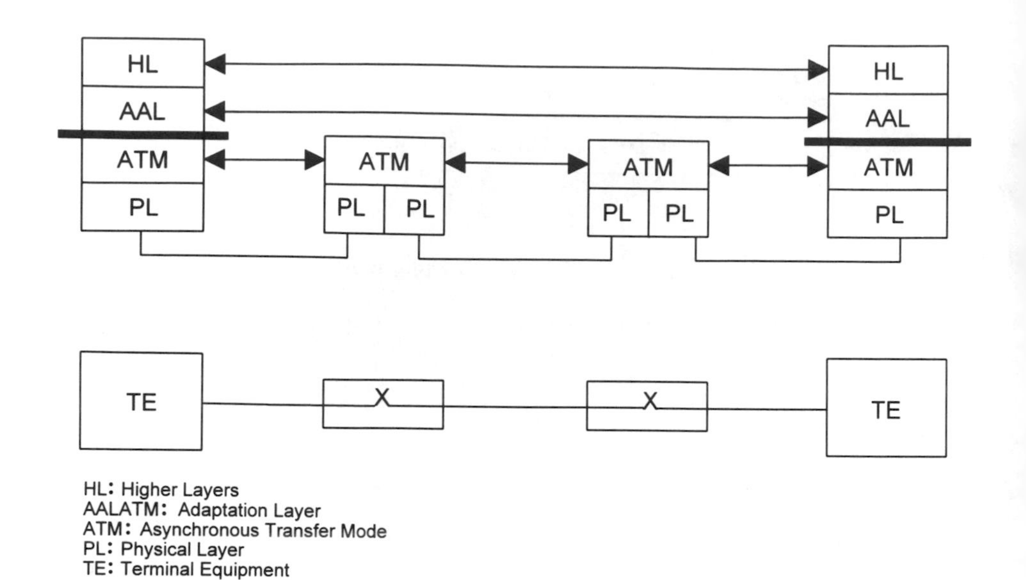

Figure 8.3 B-ISDN layered protocol communication.

may allocate resources at the start of the call to meet the required quality of service. The network may not accept the connection if the quality of service of existing connections cannot be maintained. Throughout the connection duration, the network polices the user traffic such that the traffic contract is not violated. Network resources are released at the end of the call.

The *ATM adaptation layer* (AAL) enhances the ATM layer to support the functions required by the next higher layer, in a service specific manner. AAL functions typically include:

- Error control of transmission errors in the information field;
- Segmentation and reassembly of higher layer information into ATM cells;
- Handling of cell loss and misinserted cells in the ATM layer;
- Handling of cell delay jitter;
- Transfer of timing information.

To minimize the proliferation of AAL protocols, four service classes, based upon the end-to-end timing relationship, constant or variable bit rate, and connection mode, are recognized. Class A and Class B cover real-time services. The former class is for constant bit rate services, while the latter is for variable bit rate service.

Particular AAL types under consideration for audiovisual/multimedia communications are type 1, type 2, and type 5. Methods of network adaptation that cover those layers between ATM and elementary streams, including MPEG Transport and Program Streams as part of it, are now being studied by ITU-T SG15 toward developing recommendations for audiovisual communications in the ATM environments.

8.1.1 Constant Bit Rate (CBR) and Variable Bit Rate (VBR) Coding

Restrictions of traditional circuit-switched networks have meant that all commercial digital video codecs operate at a CBR, despite the inherently varying information content of a motion video sequence (being dependent on, for example, changing image complexity, degree of motion, and frequency of scene changes). The internally varying rate in these codecs is smoothed by buffering and dynamic control of codec parameters (such as sensitivity and quantizer stepsize) to ensure that the buffer neither empties nor overflows. Such codecs operate in a fixed-rate but variable quality, mode.

ATM Networks will support VBR coded video, allowing the transmitted bit rate to dynamically reflect the information content of the changing video signal, limited by the maximum channel capacity and parameters agreed upon with the network management system. A VBR codec can therefore (usually) maintain a fixed-quality, VBR mode of operation.

8.2 ADAPTATION OF H.320 VISUAL TELEPHONE TERMINALS TO B-ISDN ENVIRONMENTS (H.321)

Recommendation H.321 describes technical specifications for adapting narrowband visual telephone terminals, as defined in Recommendation H.320, to broadband ISDN environments. The terminal conforming to this recommendation interworks with the same type of terminals (that is, other H.321 terminals) accommodated in B-ISDN as well as existing H.320 terminals accommodated in N-ISDN.

It is noted that some of the functionalities supported by H.321 terminals are also supported by broadband audiovisual terminals defined in Recommendation H.310 (see Section 8.3). The interworking among H.310, H.321, and H.320 terminals is a mandatory requirement. Interworking between H.320 and H.321 terminals is achieved since the different H.321 terminal types, defined in this recommendation, include the same functions supported by the corresponding H.320 terminal types. Interworking between H.320/H.321 and H.310 terminals is achieved through a common set of H.320/H.321 functions (defined in Recommendation H.310). For example, in addition to supporting the ITU-T Recommendation H.262 video (MPEG2 video), H.310 terminals shall support Recommendation H.261, which is part of both Recommendation H.320 and Recommendation H.321.

In H.321 terminals, the adaptation of H.320 functions over B-ISDN is achieved through AAL 1. Both *segmentation-and-reassembly* (SAR) and *convergence sublayer* (CS) functions, as defined in Recommendation I.363, are considered in this Recommendation.

H.321 terminals have the same inband functionalities as those supported by H.320 terminals, that is, as defined in Recommendations H.242, H.230, and H.221. Extra broadband-related signaling functions, such as negotiation for the use of the adaptive clock recovery method (asynchronous mode), can be accomplished through Q.2931 information elements.

A generic architecture of an H.321 terminal is shown in Figure 8.4, where constituent elements and corresponding recommendations are indicated. The figure includes such functional units as a video I/O equipment, an audio I/O equipment, a telematic equipment, a system control unit, video and audio codecs, an audio delay unit, and a mux/demux unit.

The AAL, ATM, and physical units provide the adaptation and interface functions required for accommodating an H.321 terminal over a broadband network.

8.3 HIGH-RESOLUTION BROADBAND AUDIOVISUAL COMMUNICATION SYSTEMS (H.310)

The resolution of an H.321 terminal is limited to the CIF quality specified in the H.261 standard. The purpose of the H.310 terminal is to provide a higher resolution (full frame rather than one field) capability as defined by the MPEG2 and H.262 video coding standards.

ITU Recommendation H.310 terminals are intended for the support of the following applications via the B-ISDN network:

- Conversational services;
- Retrieval services;
- Messaging services;
- Distribution services with user individual presentation control;
- Distribution services without user individual presentation control.

Figure 8.5 shows a generic H.310 broadband audiovisual communication system. It consists of terminal equipment, network, MCU, and the constituent elements of the terminal equipment. The corresponding recommendations/draft recommendations are also identified.

8.3.1 Communication Mode

Bidirectional and unidirectional (receive-only and transmit-only) audiovisual terminals are defined by this recommendation. The capabilities of H.310

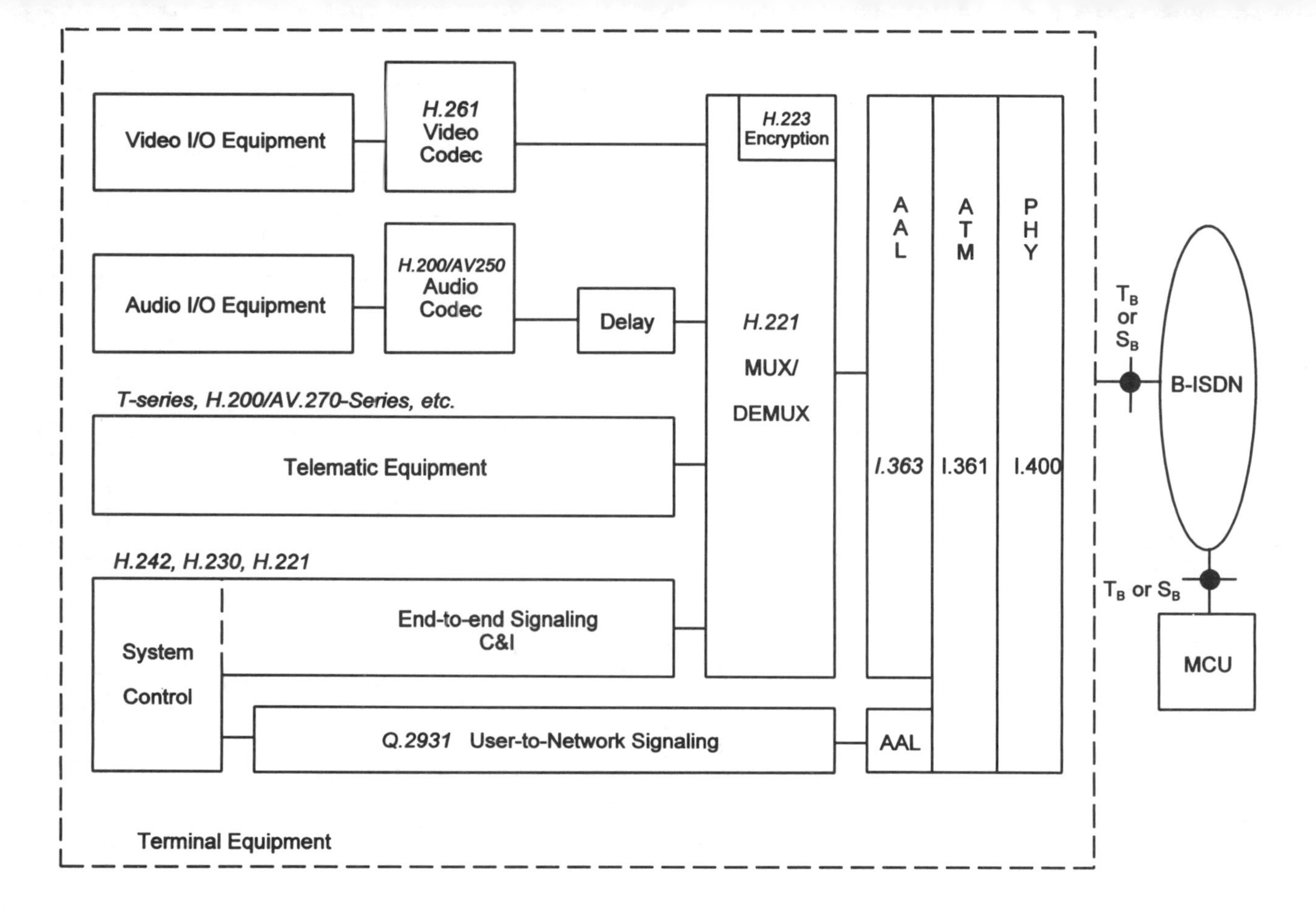

Figure 8.4 Generic architecture of an H.321 terminal.

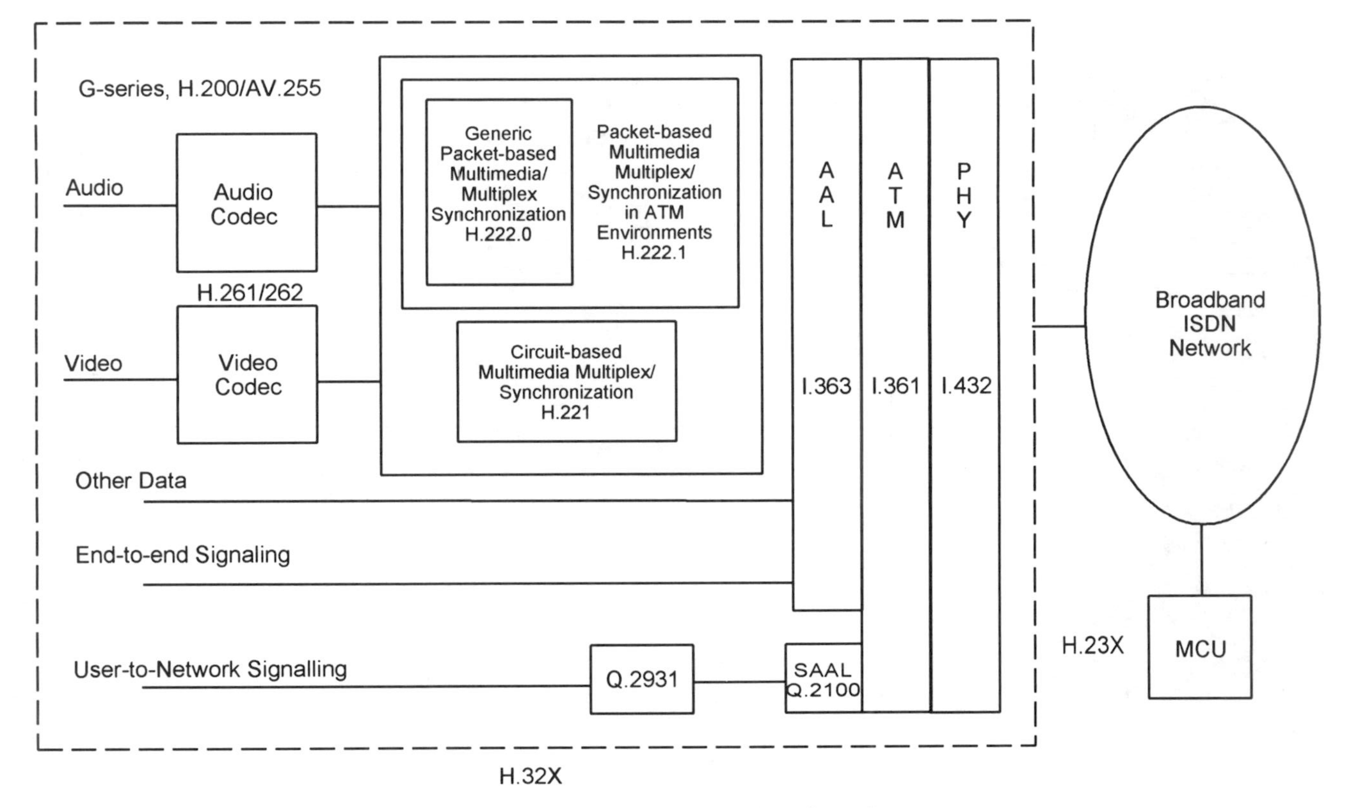

Figure 8.5 H.310 broadband audiovisual communication system and terminal configuration.

terminals are classified into five attributes: *video codec capabilities* (VCC), *audio codec capabilities* (ACC), *network adaptation capabilities* (NAC), *control & indication capabilities* (CIC), and *other data capabilities* (ODC). A communication mode is defined as a combination of those capabilities that are employed by a terminal at an instance of audiovisual communication. Since a communication session may be asymmetric for many applications in the broadband environment, these attributes are specified separately for the *transmit end* (TE) and the *receive end* (RE).

8.3.2 Video Codec Capabilities (VCC)

Both H.261 and H.262 video capabilities are considered in this recommendation. H.262 is the ITU recommendation that was developed jointly with MPEG2 (see Chapter 10). All H.310 bidirectional terminals support ITU-T Recommendation H.261 with both the CIF and QCIF picture resolutions. This enables the interworking between H.310 and a wide range of existing and future H.320/H.321 terminals. In addition, all H.310 terminals must be capable of decoding any coded video bitstream that is in conformance with Recommendation H.262 *Main Profile* (MP) at *Main Level* (ML). H.262 *MP at High Level* (MP@HL) and *MP at High*-1440 *Level* (MP@H14L) are supported by some H.310 terminals. Therefore, in addition to supporting the H.261 (CIF and QCIF) video coding modes, all H.310 bidirectional terminals support at least one of the following video standards:

- H.262 MP@ML;
- H.262 MP@H14L;
- H.262 MP@HL.

8.3.3 Audio Codec Capabilities (ACC)

Three types of audio coding standards are considered in Recommendation H.310, namely, ITU-T G-series (G.711, G.722, and G.728), ISO/IEC 1172-3 (MPEG1 audio), and ISO/IEC 13818-3 (MPEG2 audio). All bidirectional H.310 terminals must support ITU-T Recommendations G.711, G.722, and G.728. This enables existing and future H.320/H.321 terminals to interwork with H.310 terminals. In addition, all H.310 terminals should be capable of decoding any audio bitstream that is in conformance with the ISO/IEC 11172-3 (MPEG1 audio) standard. The support of the ISO/IEC 13818-3 audio standard (MPEG2 audio) is optional for some H.310 terminals.

8.3.4 Terminal Type

All H.310 terminal types can be classified into class A or B. The major difference between the two classes is their level of support for the H.262 video coding mode. Type A terminals support H.262 *MP at Main Level* (MP@ML), while type B terminals support *MP at High and High*-1440 *Levels* (MP@HL and MP@H14L). Each class is further divided into the three main terminal types *receive-only terminal* (ROT), *sending-only terminal* (SOT), and bidirectional *receive-and-send terminal* (RAST). Each of these terminal types support different mandatory and optional H.310 capabilities.

8.3.5 Intercommunication With N-ISDN Terminals

In general, it is desirable that bidirectional H.310 terminals be able to interact with H.320/H.321 terminals. In an intercommunication session with an H.320/ H.321 terminal, bidirectional H.310 terminals function as H.321 terminals. Support of B, 2B, and H0 communication modes of H.320 in bidirectional H.310 terminals is mandatory. Other communication modes (for example, H.11 and H.12) are optional.

8.4 VISUAL TELEPHONE SYSTEMS FOR LOCAL AREA NETWORKS THAT PROVIDE A GUARANTEED QUALITY OF SERVICE (H.322)

Recommendation H.322 covers the technical requirements for visual telephone services defined in H.200/AV.120-Series Recommendations in those situations where the transmission path includes one or more LANs, each of which is configured and managed to provide a guaranteed *Quality of Service* (QoS) equivalent to that of N-ISDN such that no additional protection or recovery mechanisms beyond those mandated by Recommendation H.320 need be provided in the terminals. Pertinent parameters are the data error and loss properties and variation of transit delay. H.322 also requires that the ISDN clock is available at the terminals. An example of a suitable LAN is IEEE 802.9a isochronous services with *carrier sense multiple access with collision detection* (CSMA/ CD) *media access control* (MAC) service. This LAN connection is known generally as ISO-Ethernet. An ISO-Ethernet LAN consists of a star network where each user connection to the hub is essentially an ISDN channel that can handle up to 96 B channels (6.1 Mbps). Since the quality of service on each of the user connections is guaranteed and ISDN-like, it is clear why the H.322 Recommendation is a simple extension of H.320.

Recommendation H.323 addresses the use of some other LANs that are unable to provide the underlying performance assumed by this H.322 Recommendation (see Section 8.5). Although some LANs offer satisfactory QoS to support Recommendation H.322, they do not themselves provide standardized means of identifying the intended called terminal. For such LANs the procedures for terminal addressing defined in Recommendation H.323 shall apply in addition to the provisions of this H.322 Recommendation (see Section 8.5).

The H.322 Recommendation does not encompass ATM LANs because they are within the scope of Recommendation H.321.

Systems and terminal equipment complying with Recommendation H.322 are able to interwork with each other and with those complying with Recommendations H.320, H.321, and H.323. The concept is shown in Figure 8.6 in which any terminal can be connected to any other. The H.322 gateway unit provides an interconnection between the LAN and the *wide area network* (WAN), which may be N-ISDN or B-ISDN or both. An H.322 terminal communicates with another H.322 terminal on the same LAN directly. The gateway may be connected via N-ISDN or B-ISDN to other gateways and LANs to provide communication between H.322 or H.323 terminals that are not on the same LAN.

Although Recommendation H.322 specifically addresses visual telephone systems, the methods used do not depend on the content of the signals carried. Consequently, this recommendation has more general applicability to connecting terminals, originally designed for N-ISDN, over LAN or hybrid LAN and ISDN networks.

Figure 8.7 is a block diagram of the Recommendation H.322 terminal. All elements are identical to those specified in the Recommendation except for the LAN interface.

8.5 VISUAL TELEPHONE SYSTEM FOR LANs THAT PROVIDE A NONGUARANTEED QUALITY OF SERVICE (H.323)

8.5.1 Scope

ITU Recommendation H.323 covers the technical requirements for narrowband visual telephone services defined in H.200/AV.120-Series Recommendations in those situations where the transmission path includes one or more LANs,

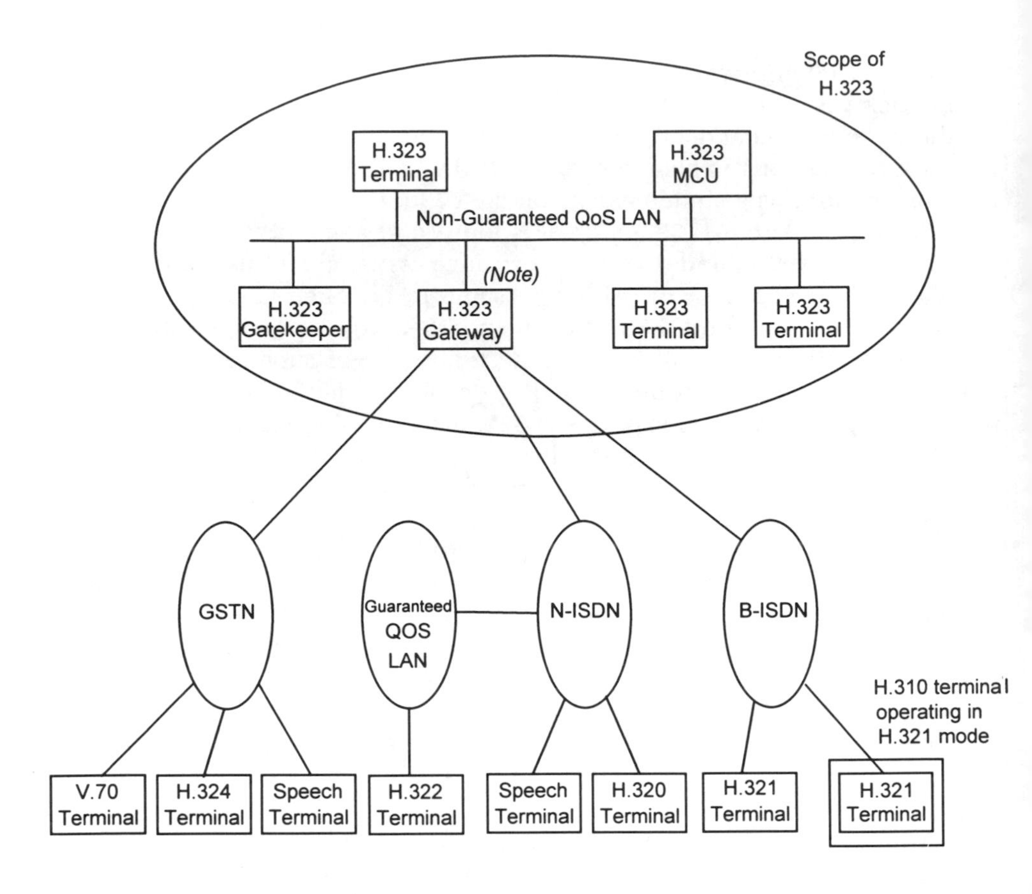

Figure 8.6 Interoperability of H.323 terminals. (Note: A gateway may support one or more of the GSTN, N-ISDN, and/or B-ISDN connections.)

which may not provide a guaranteed QoS equivalent to that of N-ISDN. Examples of this type of LAN are:

- Ethernet (IEEE 802.3);
- Fast Ethernet (IEEE 802.10);
- FDDI (nonguaranteed QoS mode);
- Token Ring (IEEE 802.5).

H.323 terminals may be used in multipoint configurations and may interwork with H.310 terminals on B-ISDN, H.320 terminals on N-ISDN, H.321 terminals on B-ISDN, H.322 terminals on guaranteed QoS LANs, H.324 terminals on GSTN and wireless networks, and V.70 terminals on GSTN. See Figure 8.6.

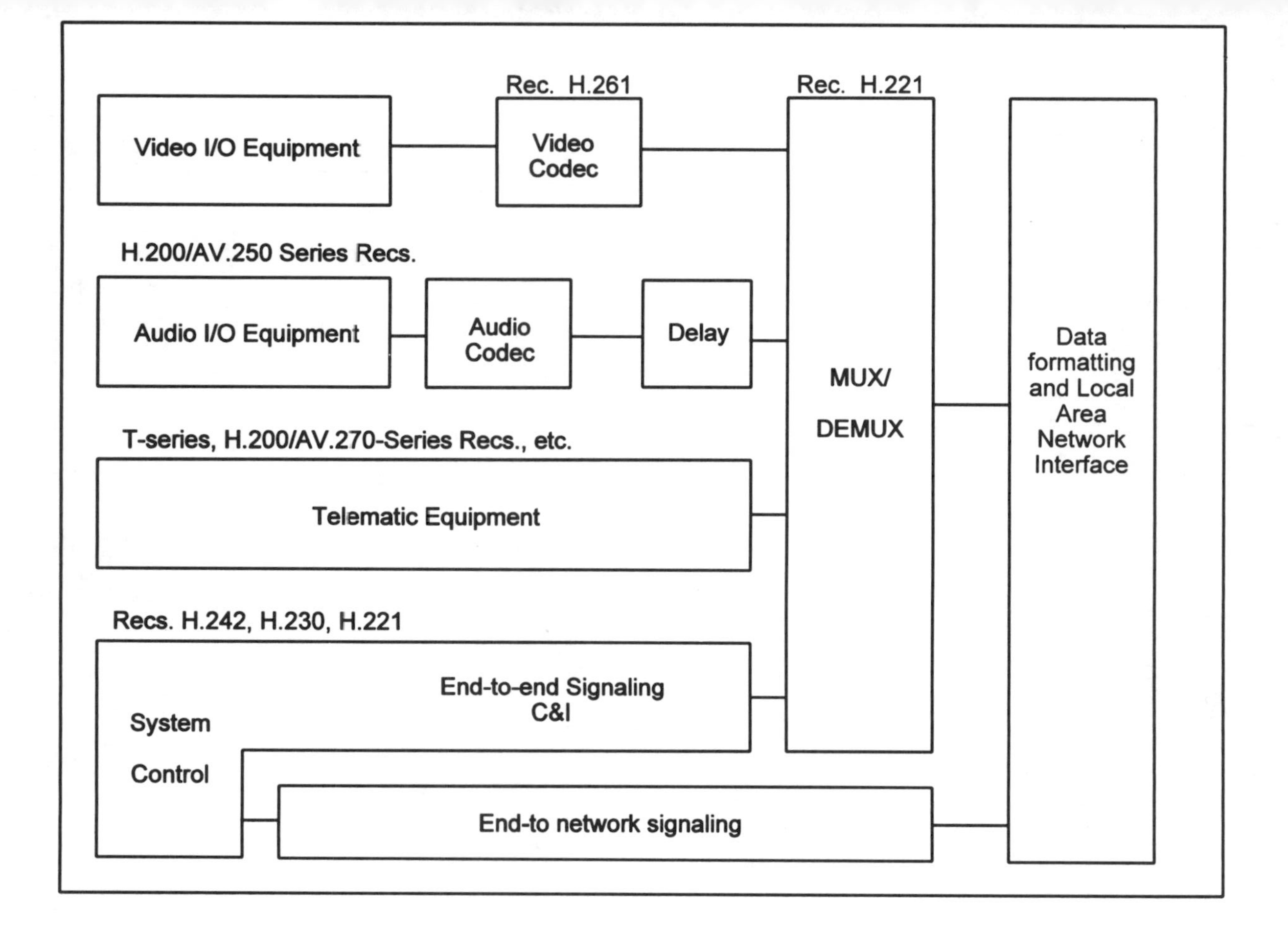

Figure 8.7 Block diagram of H.322 terminal elements.

The scope of H.323 does not include the LAN itself or the transport layer that may be used to connect various LANs. Only elements needed for interaction with the *switched circuit network* (SCN) are within the scope of H.323. The combination of the H.323 Gateway, the H.323 terminal, and the out-of-scope LAN appears on the SCN as an H.320, H.310, H.324, or V.70 terminal.

This recommendation describes the components of an H.323 system, which includes terminals, gateways, gatekeepers, multipoint controllers, multipoint processors, and MCUs. Control messages and procedures within this recommendation define how these components communicate.

The components described in this recommendation consist of H.323 endpoints and H.323 entities. The endpoints can call and are callable according to the call setup procedures. The entities are not callable; however, they can be addressed to perform their specific functions. For example, a terminal cannot place a call to a gatekeeper, however, the gatekeeper is addressed as part of the call establishment procedures.

8.5.2 Terminal Characteristics

An example of an H.323 terminal is shown in Figure 8.8. The diagram shows the user equipment interfaces, video codec, audio codec, telematic equipment,

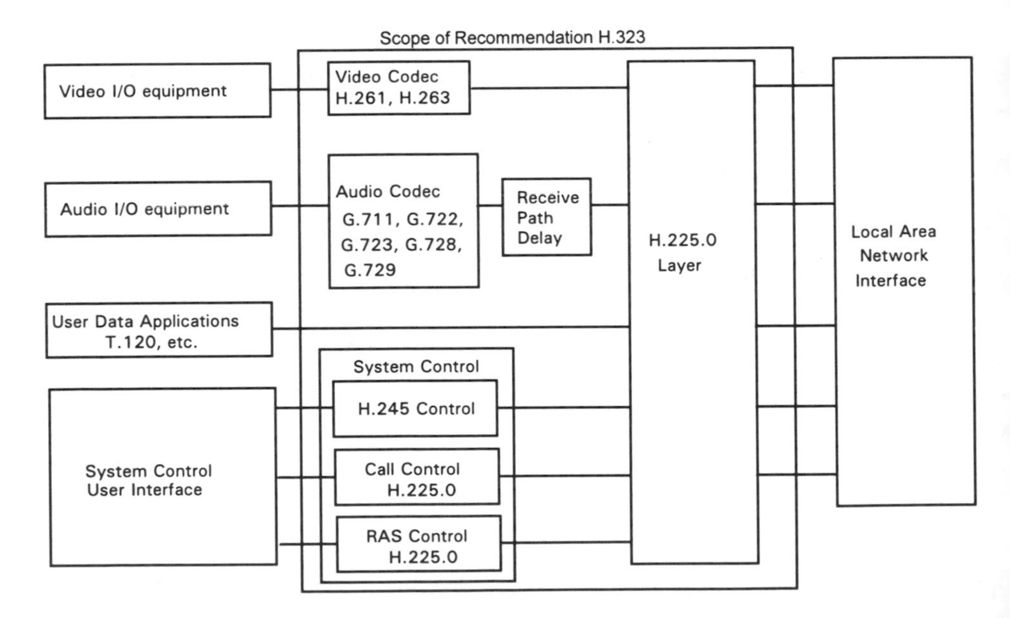

Figure 8.8 H.323 terminal equipment.

H.225.0 layer, system control functions, and the interface to the LAN. All H.323 terminals shall have a system control unit, H.225.0 layer, network interface, and an audio codec unit. The video codec unit and user data applications are optional.

The following elements are not within the scope of H.323 and are therefore not defined within this recommendation:

- Attached audio devices providing voice activation sensing, microphone and loudspeaker, telephone instrument or equivalent, multiple microphones mixers, and acoustic echo cancellation;
- Attached video equipment providing cameras and monitors, their control and selection, and video processing to improve compression or provide split screen functions;
- Data applications and associated user interfaces that use T.120 or other data services over the data channel.
- Attached network interface, which provides the interface to the LAN, supporting appropriate signaling and voltage levels, in accordance with national and international standards.

The following elements are within the scope of H.323 and are therefore subject to standardization and are defined within this recommendation.

- The video codec (for example, H.261) encodes the video from the video source (that is, camera) for transmission and decodes the received video code that is output to a video display.
- The audio codec (for example, G.711) encodes the audio signal from the microphone for transmission and decodes the received audio code that is output to the loudspeaker.
- The data channel supports telematic applications such as electronic whiteboards, still image transfer, file exchange, database access, and audiographics conferencing. The standardized data application for real-time audiographics conferencing is T.120. Other applications and protocols may also be used via H.245 negotiation as specified in Section 6.2.7.
- The system control unit (H.245, H.225.0) provides signaling for proper operation of the H.323 terminal. It provides for call control, capability exchange, signâling of commands and indications, and messages to open and fully describe the content of logical channels.
- H.225.0 layer (H.225.0) formats the transmitted video, audio, data, and control streams into messages for output to the network interface and retrieves the received video, audio, data, and control streams from messages that have been input from the network interface. In addition, it

performs logical framing, sequence numbering, error detection, and error correction as appropriate to each media type.

The LAN interface is implementation specific and is outside the scope of this recommendation. However, the LAN interface shall provide the services described in Recommendation H.225.0. This includes that reliable (for example TCP, SPX) end-to-end service is mandatory for the H.245 Control Channel, the data channels, and the call signaling channel. Unreliable (for example UDP, IPX) end-to end service is mandatory for the audio channels, the video channels, and the RAS channel. These services may be duplex or simplex, unicast or multicast depending on the application, the capabilities of the terminals, and the configuration of the LAN.

The video codec is optional. All H.323 terminals providing video communications shall be capable of encoding and decoding video according to H.261 QCIF. Optionally, a terminal may also be capable of encoding and decoding video according H.261 CIF or H.263 SQCIF, QCIF, CIF, 4CIF, and 16CIF. If a terminal supports H.263 with CIF or higher resolution, it shall also support H.261 CIF. All terminals that support H.263 shall support H.263 QCIF. The H.261 and H.263 codecs, on the LAN, shall be used without BCH error correction and without error correction framing.

Other video codecs and other picture formats may also be used via H.245 negotiation. More than one video channel may be transmitted and/or received, as negotiated via the H.245 control channel. The H.323 terminal may optionally send more than one video channel at the same time, for example, to convey the speaker and a second video source. The H.323 terminal may optionally receive more than one video channel at the same time, for example, to display multiple participants in a distributed multipoint conference.

CIF and QCIF are defined in H.261. SQCIF, 4CIF, and 16CIF are defined in H.263. For the H.261 algorithm, SQCIF is any active picture size less than QCIF, filled out by a black border, and coded in the QCIF format. For all these formats, the pixel aspect ratio is the same as that of the CIF format.[1]

The video bitrate, picture format, and algorithm options that can be accepted by the decoder are defined during the capability exchange using H.245. The encoder is free to transmit anything that is within the decoder capability set. The decoder should have the possibility to generate requests via H.245 for a certain mode, but the encoder is allowed to simply ignore these requests if they are not mandatory modes. Decoders that indicate capability for a particular algorithm option will also be capable of accepting video bitstreams that do not make use of that option.

[1]The resulting picture aspect ratio for H.263 SQCIF is different from the other formats.

H.323 terminals will be capable of operating in asymmetric video bit rates, frame rates, and, if more than one picture resolution is supported, picture resolutions. For example, this will allow a CIF capable terminal to transmit QCIF while receiving CIF pictures.

All H.323 terminals shall have an audio codec. All H.323 terminals shall be capable of encoding and decoding speech according to Recommendation G.711. All terminals shall be capable of transmitting and receiving A-law and mu-law. A terminal may optionally be capable of encoding and decoding speech using Recommendations G.722, G.728, G.729, MPEG1 audio, and G.723.1. The audio algorithm used by the encoder shall be derived during the capability exchange using H.245. The H.323 terminal should be capable of asymmetric operation for all the audio capabilities that it has declared within the same capability set; for example, it should be able to send G.711 and receive G.728 if it is capable of both.

The audio stream is formatted as described in H.225.0. Each logical channel opened for audio shall be accompanied by a logical channel opened for audio control. The audio control channel is described in H.225.0.

The H.323 terminal may optionally send more than one audio channel at the same time, for example, to allow two languages to be conveyed.

One or more data channels are optional. The data channel may be unidirectional or bidirectional depending on the requirements of the data application.

T.120 is the default basis of data interoperability between an H.323 terminal and other H.323, H.320, or H.310 terminals. Where any optional data application is implemented using one or more of the ITU-T recommendations that can be negotiated via H.245, the equivalent T.120 application, if any, will be one of those provided. A terminal that provides far-end camera control using H.281 and H.224 is not required also to support a T.120 far-end camera control protocol.

The H.245 control function uses the H.245 control channel to carry end-to-end control messages governing operation of the H.323 entity, including capabilities exchange, opening and closing of logical channels, mode preference requests, flow control messages, and general commands and indications.

H.245 signaling is established between two endpoints, an endpoint and an MC, or an endpoint and a gatekeeper. The endpoint shall establish exactly one H.245 control channel in each direction for each call in which the endpoint is participating. This channel shall use the messages and procedures of Recommendation H.245. Note that a terminal, MCU, gateway, or gatekeeper may support many calls and, thus, many H.245 control channels. The H.245 control channel shall be carried on logical channel 0. Logical channel 0 will be considered to be permanently open from the establishment of the H.245 control channel until the termination of this channel. The normal procedures for opening and closing logical channels will not apply to the H.245 control channel.

H.245 specifies a number of independent protocol entities that support endpoint-to-endpoint signaling. A protocol entity is specified by its syntax (messages), semantics, and a set of procedures that specify the exchange of messages and the interaction with the user. H.323 endpoints will support the syntax, semantics, and procedures of the following protocol entities:

- Master/slave determination;
- Capability exchange;
- Logical channel signaling;
- Bidirectional logical channel signaling;
- Close logical channel signaling;
- Mode request;
- Round trip delay determination;
- Maintenance loop signaling.

General commands and indications will be chosen from the message set contained in H.245. In addition, other command and indication signals may be sent that have been specifically defined to be transferred inband within video, audio, or data streams (see the appropriate recommendation to determine if such signals have been defined).

8.5.3 Definitions

The following are definitions for terms used in this chapter.

- *Call.* Point-to-point multimedia communication between two H.323 endpoints. The call begins with the call setup procedure and ends with the call termination procedure. The call consists of the collection of reliable and unreliable channels between the endpoints. In case of interworking with some SCN endpoints via a gateway, all the channels terminate at the gateway where they are converted to the appropriate representation for the SCN end system.
- *Endpoint.* An H.323 terminal, gateway, or MCU. An endpoint can call and be called. It generates and/or terminates information streams.
- *Gatekeeper.* The *gatekeeper* (GK) is an H.323 entity on the LAN that provides address translation and controls access to the local area network for H.323 terminals, gateways, and MCUs. The gatekeeper may also provide other services to the terminals, gateways, and MCUs such as bandwidth management and locating gateways.
- *Gateway.* An H.323 *gateway* (GW) is an endpoint on the local area network that provides for real-time, two-way communications between H.323 terminals on the LAN and other ITU terminals on a WAN or to another H.323

gateway. Other ITU terminals include those complying with Recommendations H.310 (H.320 on B-ISDN), H.320 (ISDN), H.321 (ATM), H.322 (GQOS-LAN), H.324 (GSTN), H.324M (Mobile), and V.70 (DSVD).

- *H.245 control channel.* Reliable channel used to carry the H.245 control information messages (following H.245) between two H.323 endpoints.
- *H.245 logical channel.* Channel used to carry the information streams between two H.323 endpoints. These channels are established following the H.245 OpenLogicalChannel procedures. An unreliable channel is used for audio, audio control, video, and video control information streams. A reliable channel is used for data and H.245 control information streams. There is no relationship between a logical channel and a physical channel.
- *Local area network (LAN).* A shared or switched medium, peer-to-peer communications network that broadcasts information for all stations to receive within a moderate-sized geographic area, such as a single office building or a campus. The network is generally owned, used, and operated by a single organization. In the context of H.323, LANs also include internetworks composed of several LANs that are interconnected by bridges or routers.
- *Terminal.* An H.323 terminal is an endpoint on the local area network that provides for real-time, two-way communications with another H.323 terminal, gateway, or MCU. This communication consists of control, indications, audio, moving color video pictures, and/or data between the two terminals. A terminal may provide speech only, speech and data, speech and video, or speech, data, and video.
- *Zone.* A zone is the collection of all *terminals* (Tx), GW, and MCUs managed by a single GK. A zone includes at least one terminal and may or may not include gateways or MCUs. A zone has one and only one GK. A zone may be independent of LAN topology and may be comprised of multiple LAN segments that are connected using *routers* (R) or other devices.

Multipoint Graphic Communications (T.120)

9

It is recognized that in many situations it is not necessary to transmit a motion video signal to achieve a satisfactory electronic conferencing capability. Instead the conference may require only audio and graphics. This chapter describes a new ITU standard, namely, T.120, which provides the ability to interactively exchange graphics (or any data) on a multipoint basis. The audio portion of the conference would be provided independently from the T.120 process.

Traditionally, telephony services have been confined to point-to-point operation. To support group activities such as meetings and conferences involving physically separated participants, there is a requirement to join together more than two locations. The term multipoint communication simply describes the interconnection of multiple terminals as shown in Figure 9.1. Normally, a special network element, known as a multipoint control unit (MCU), or simply a bridge, is required to provide this function. The raw communication data

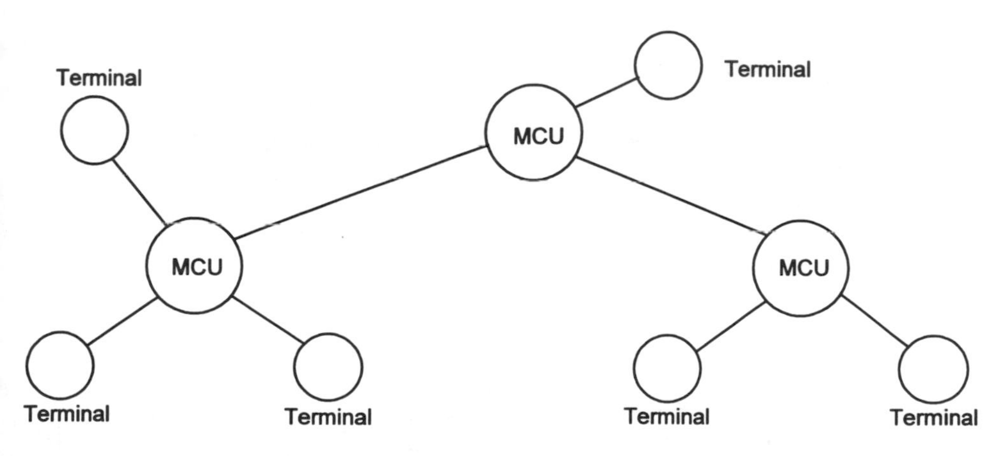

Figure 9.1 Typical multipoint configuration.

stream would typically consist of one or more of the three media elements: audio, video, and data. The term multimedia is now widely used to collectively describe this grouping. The T.120 series of recommendations applies to the data element, which would typically be used both to provide a data communications service and management of any other media services present.

The T.120 ITU Recommendation introduces the T.12x series of standards, collectively referred to as the *T.120 series*. The T.120 protocol is a means of telecommunicating all forms of data/telematic media between two or more multimedia terminals and of managing such communication. It can also manage real-time conversational speech and video whose information signals are transmitted on channels separate from that carrying the T.120 protocol.

The T.120 model is composed of a communications infrastructure and application protocols that make use of it. Figure 9.2 shows the full model with both standardized and nonstandardized components. The model serves both to show the scope of the T.120 suite of recommendations and the relationship between each of the recommendations and other elements in the system. Generally, each layer provides services to the layer above and communicates to its peer(s) by sending *protocol data units* (PDUs) via services provided by the layer below. This discussion will address each of the major functional levels in Figure 9.2, that is, application protocols, node controller, and communications infrastructure.

9.1 NODE CONTROLLER

In a T.120-based system the term *node controller* is used to describe the management function or role at a terminal or MCU. It is the node controller that issues the primitives to GCC that start and control the communication session. The node control function will exist in all systems, providing the interface above GCC and performing any other required management functions. The node controller itself is outside the scope of the T.120 recommendations, and only where it communicates to GCC and any other standardized elements are the interfaces defined (by association).

9.2 COMMUNICATIONS INFRASTRUCTURE

The communications infrastructure, illustrated in Figure 9.2 , provides simultaneous multipoint connectivity with reliable data delivery. It can accommodate multiple independent applications concurrently using the same multipoint environment. Connections can be any combination of circuit-switched telecommunications networks and packet-based LANs and data networks. It is composed of three standardized components: *generic conference control* (GCC), the

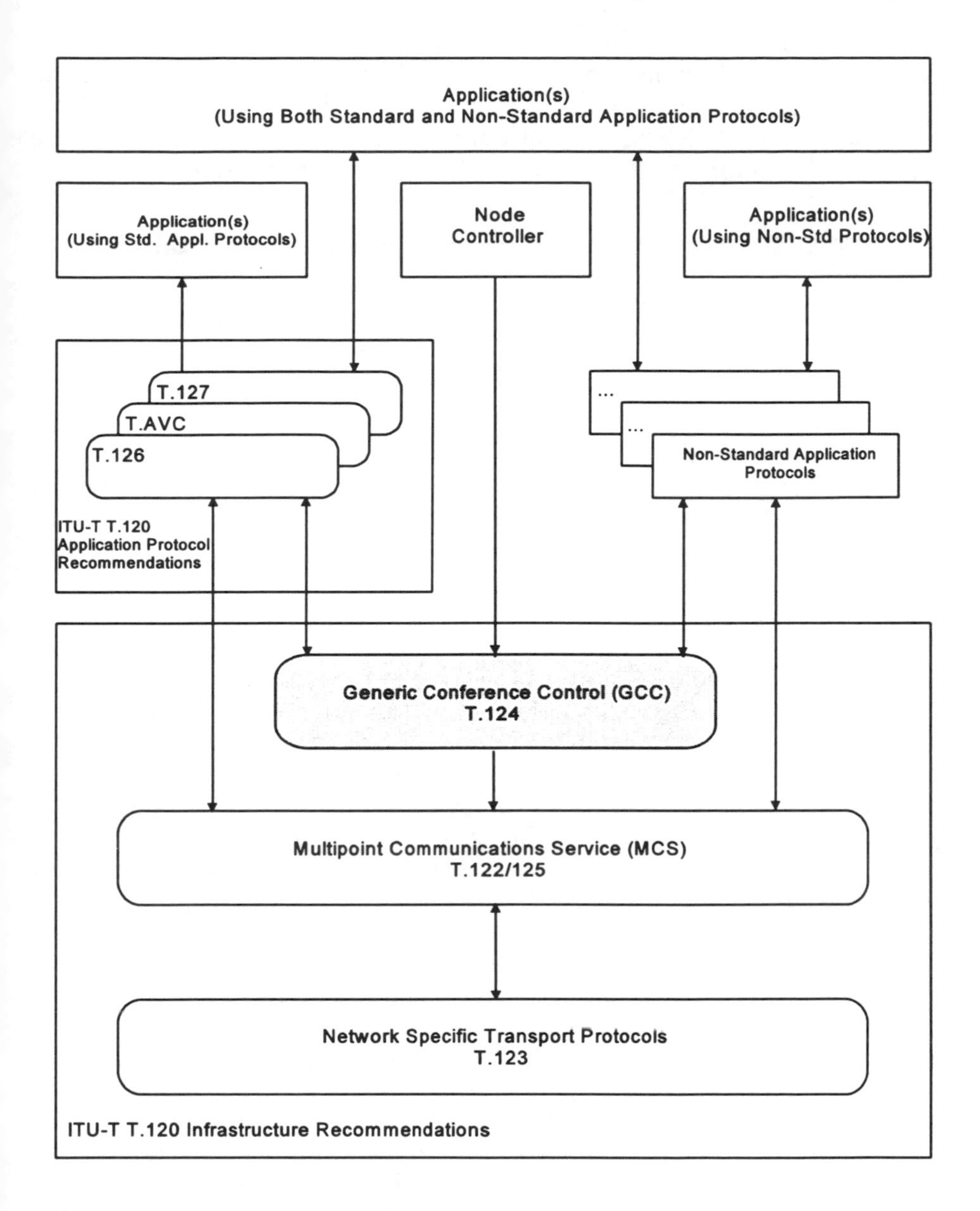

Figure 9.2 The T.120 series of recommendations.

multipoint communications service (MCS), and transport protocol profiles for each of the supported networks. Adding GCC provides a range of facilities oriented to the use of MCS in an electronic meeting or conference but believed to be generally useful for other multipoint communications requirements.

9.2.1 T.122/125 Multipoint Communications Service (MCS)

T.122 defines the multipoint services available to the developer, while T.125 specifies the data transmission protocol. Together they form MCS, the multipoint "engine" of the T.120 conference. MCS relies on T.123 to actually deliver the data. (Use of MCS is entirely independent of the actual T.123 transport stack(s) that is loaded.)

MCS is a powerful tool that can be used to solve virtually any multipoint application design requirement. MCS is an abstraction of a rather complex organism. Learning to use MCS effectively is the key to successfully developing real-time applications.

9.2.1.1 How MCS Works

In a conference, multiple endpoints (or MCS nodes) are logically connected to form what T.120 refers to as a domain. Domains generally equate to the concept of a conference. Applications may actually be attached to multiple domains simultaneously. For example, the chairperson of a large online conference may simultaneously monitor information being discussed among several activity groups. If the chairperson wanted to bring two of these groups together to share ideas, the conference provider could use the sophisticated domain merge facility to accomplish this request.

In a T.120 conference, nodes connect upward to a MCU. The MCU model in T.120 provides a reliable approach that works well in both public and private networks. Multiple MCUs may be easily connected in a single domain. Figure 9.3 illustrates potential topology structures. Each domain has a single top provider or MCU that houses the information base that is critical to the conference. If the top provider either fails or leaves a conference, the conference is terminated. If a lower level MCU (that is, not the top provider) fails, only the nodes on the tree below that MCU are dropped from the conference.

One of the critical features of the T.120 approach is the ability to direct data. This capability allows applications to communicate very efficiently. MCS applications direct data within a domain via the use of channels. While an application may only send data along one channel at a time, it may simultaneously subscribe or receive information from many channels at once. These channel assignments can be dynamically changed during the life of the conference.

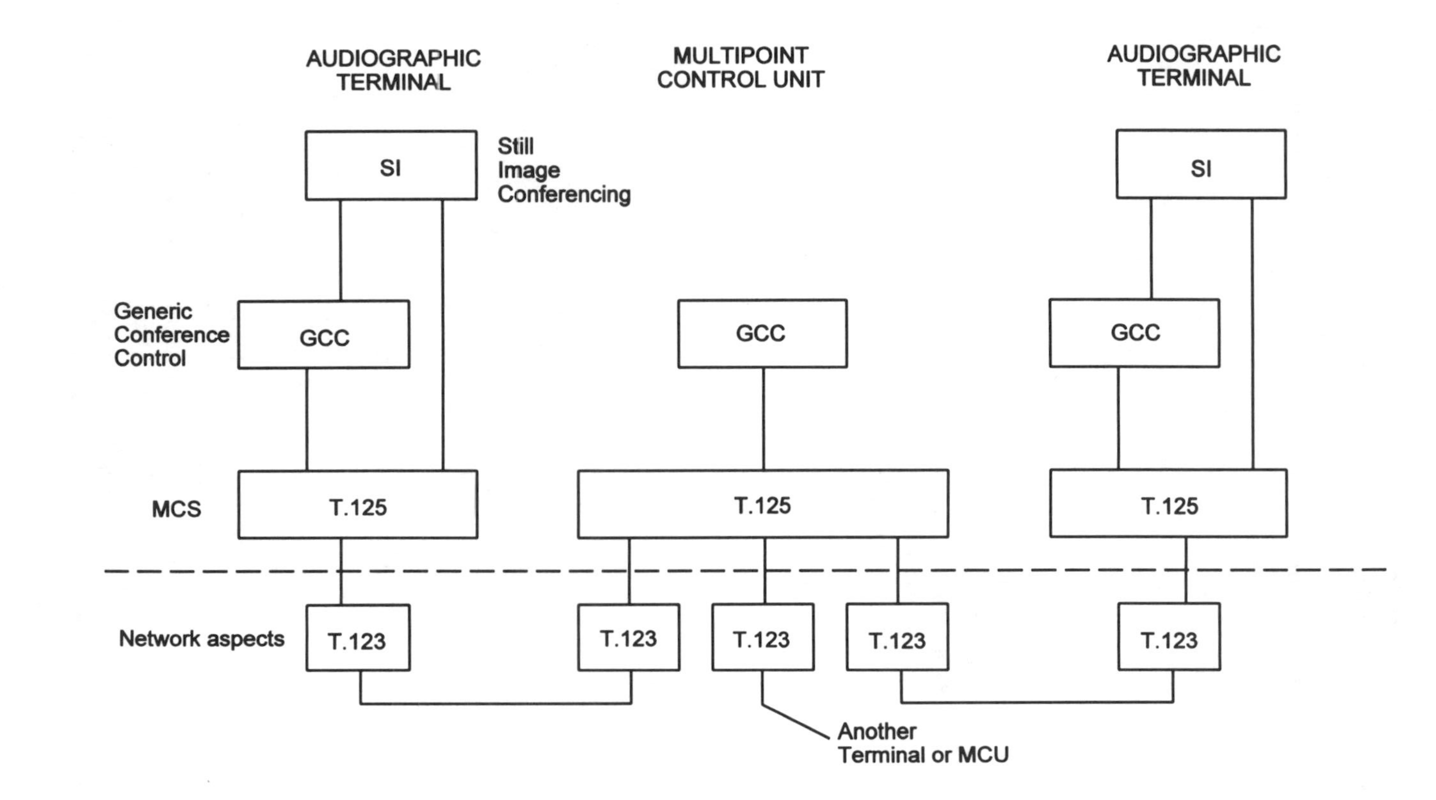

Figure 9.3 Framework of the AGC protocol suite.

It is up to the application designer to determine how to use channels within an application. For example, an application may send control information along a single channel and application data along a series of channels that may vary depending upon the type of data being sent. The application developer may also take advantage of the MCS concept of private channels to direct data to a discrete subset of a given conference.

Data may be sent with one of four priority levels. MCS applications may also specify that data is routed along the quickest path of delivery using the standard send command. If the application uses the uniform send command, it ensures that data from multiple senders will arrive at all destinations in the same order. Uniform data always travels all the way up the tree to the top provider.

There are no constraints on the size of the data sent from an application to MCS. Segmentation of data is automatically performed on behalf of the application. However, it is the application's responsibility to reassemble the data upon receiving it by monitoring flags provided when the data is delivered.

Tokens are the last major facility provided by MCS. Services are provided to grab, pass, inhibit, release, and query tokens. Token resources may be used as either exclusive (that is, locking) or nonexclusive entities.

Tokens can be used by an application in a number of ways. For example, an application may specify that only the holder of a specific token (that is, the conductor) may send information in the conference. Another popular use of tokens is to coordinate tasks within a domain. For example, suppose a teacher wants to be sure that every student in a distance learning session answered a particular question before displaying the answer. Each node in the underlying application inhibits a specific token after receiving the request to answer the question. The token is released by each node when an answer is provided. In the background, the teacher's application continuously polls the state of the token. When all nodes have released the token, the application presents the teacher with a visual cue that the class is ready for the answer.

9.2.2 T.124 Generic Conference Control (GCC)

Generic conference control provides a comprehensive set of facilities for establishing and managing the multipoint conference. It is with GCC that we first see features that are specific to the electronic meeting.

At the heart of GCC is an important information base about the state of the various conferences it may be servicing. One node, which may be the MCU itself, serves as the top provider for GCC information. Any actions or requests from lower GCC nodes ultimately filter up to this top provider.

Using mechanisms in GCC, applications create conferences, join conferences, and invite others to conferences. As endpoints join and leave conferences, the information base in GCC is updated and can be used to automatically notify all endpoints when these actions occur. GCC also knows who is the top provider for the conference. However, GCC does not contain detailed topology information about the means by which nodes from lower branches are connected to the conference.

Every application in a conference must register its unique application key with GCC. This enables any subsequent joining nodes to find compatible applications. Furthermore, GCC provides robust facilities for applications to exchange capabilities and arbitrary feature sets. In this way, applications from different vendors can readily establish whether or not they can interoperate and at what feature level. This arbitration facility is the mechanism used to ensure backward compatibility between different versions of the same application.

GCC also provides conferences security. This allows applications to incorporate password protection or *lock* facilities to prevent uninvited users from joining a conference.

Another key function of GCC is its ability to dynamically track MCS resources. Since multiple applications can use MCS at the same time, applications rely on GCC to prevent conflicts for MCS resources, such as channels and tokens. This ensures that applications do not step on each other by attaching to the same channel or requesting a token already in use by another application.

Finally, GCC provides some capabilities for supporting the concept of conductorship in a conference. GCC provides applications with information about who the meeting conductor is and a means to transfer the conductor's "baton." The developer is free to decide how to use these conductorship facilities within the application.

9.2.3 T.123 Transport Protocol Stack Profiles

MCS expects its underlying transport to provide reliable point-to-point data delivery of its PDUs and to segment and sequence that data if necessary. T.123 (Figure 9.4) is designed to provide open and easily extended network support for both standardized and nonstandardized protocols. The basic T.123 presents a uniform OSI transport interface and services (X.214/224) to the MCS layer above. Connection-oriented profiles are provided for switched telecom and packet-switched networks. For computer networks that are not connection oriented, the profiles include additional transport layer capability to make them functionally connection oriented.

The T.120 suite provides for operation over the following networks:

- Public switched telephone network (PSTN), or compatible service with 3.1-Khz bandwidth channels using V series modems;

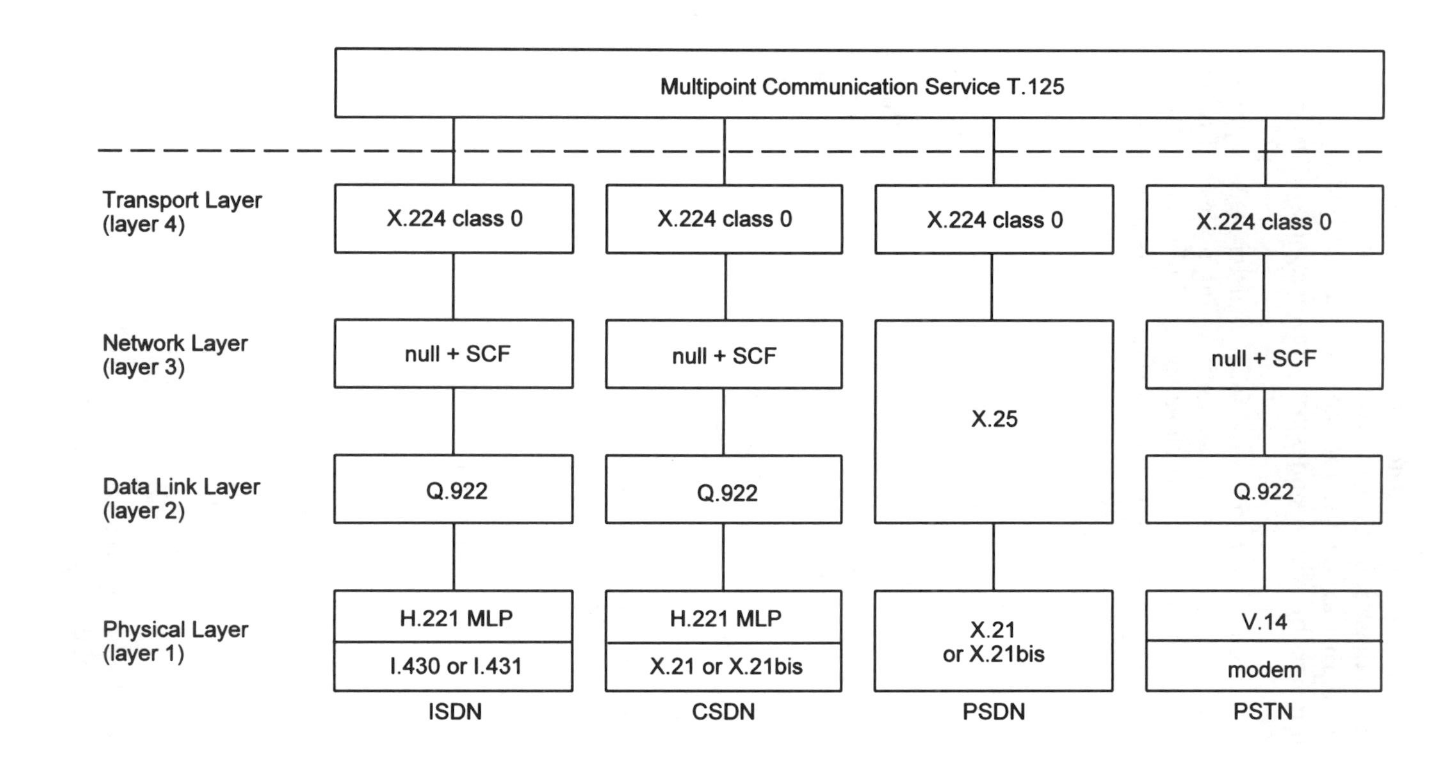

Figure 9.4 Basic mode profiles general structure.

- Integrated services digital network (ISDN) as defined in ITU-T I series recommendations;
- Packet-switched data network (PSDN) using X.25;
- Other (switched or permanent) digital circuits using H.221 framed signals;
- Use of T.120 on other networks such as LANs and ATM is currently under study.

9.3 APPLICATION PROTOCOLS

Application protocols comprise a set of PDUs and associated actions for application peer-to-peer(s) communication. These may be proprietary protocols, or they may be standardized by the ITU-T or other international or national standards bodies. The T.120 series includes a set of application protocols designed to support common facilities for multipoint communication. These protocols define and mandate a minimum requirement to ensure interworking between different implementations and include facilities for simultaneous multipoint file transfer (T.127) and audiographics protocols for still image viewing and annotation and application sharing and fax (all provided in T.126).

9.3.1 Recommendation T.126: Still Image Exchange and Annotation (SI)

T.126 defines a protocol for viewing and annotating still images transmitted between two or more applications. This capability is often referred to as document conferencing or shared whiteboarding.

An important benefit of T.126 is that it readily shares visual information between applications that are running on dramatically different platforms. For example, a Windows-based desktop application could easily interoperate with a collaboration program running on a Power Mac. Similarly, a group-oriented conferencing system, without a PC-style interface, could share data with multiple users running common PC desktop software.

T.126 presents the concept of shared virtual workspaces that are manipulated by the endpoint applications. Each workspace may contain a collection of objects that include bitmap images and annotation primitives, such as rectangles and freehand lines. Bitmaps typically originate from application information, such as a word processing document or a presentation slide. Because of their size, bitmaps are often compressed to improve performance over lower-speed communication links.

T.126 is designed to provide a minimum set of capabilities required to share information between disparate applications. Several important concepts commonly found in shared whiteboarding applications, such as support for non-bitmap data (objects), is not supported in T.126. Additionally, because T.126 is simply a protocol, it does not provide any of the API-level structures that allow application developers to easily incorporate shared whiteboarding into an application.

9.3.2 Recommendation T.127: Multipoint Binary File Transfer Protocol

This recommendation defines a protocol to support the interchange of binary files within an interactive conferencing or group working environment where the T.120 series of standards is in use. It provides mechanisms that facilitate distribution and retrieval of one or more files simultaneously. T.127 uses services provided by T.122 (MCS) and T.124 (GCC). T.127 adheres to the header structure defined in T.434.

10 ISO Audiovisual Standards (MPEG, JPEG, JBIG)

The ISO standard organization has established a working group (ISO/IECJTCI/SC2/WG11), known as the *Motion Picture Experts Group* (MPEG), to develop standards for coding audio visual (AV) signals.

The MPEG work is organized in three separate projects—MPEG1, MPEG2, and MPEG4—which are summarized below and discussed in the following sections.[1]

- MPEG1: Coding of Moving Pictures and Associated Audio for Digital Storage Media at up to About 1.5 Mbps

Part 1: Systems	IS	11172-1 (1993)
Part 2: Video	IS	11172-2 (1993)
Part 3: Audio	IS	11172-3 (1993)

- MPEG2: Generic Coding of Moving Pictures and Associated Audio Information

Part 1: Systems	IS	13818-1 (1995)
Part 2: Video	IS	13818-2 (1995)
Part 3: Audio	IS	13818-3 (1995)

- MPEG4: Very Low Bit Rate Audio-Visual Coding

Part 1: Systems	WD	14496-1
Part 2: Video	WD	14496-2
Part 3: Audio	WD	14496-3

ISO has also established two groups for coding still pictures. The *Joint Photographic Experts Group* (JPEG) works on continuous tone pictures, while the Joint Binary Imagery Group (JBIG) develops coding systems for black-white pictures. Work by these two groups will also be discussed.

[1]IS, International Standard; DIS, Draft International Standard; WD, Working Draft.

10.1 MPEG1

10.1.1 System Layer (11172-1)

An ISO 11172 bit stream is constructed in two layers: the outer-most layer is the system layer, and the inner-most is the compression layer. The system layer provides the functions necessary for using one or more compressed data streams in a system. The video and audio parts of this specification define the compression encoding layer for audio and video data. Coding of other types of data are not defined by the specification but are supported by the system layer providing that the other types of data adhere to the system constraints. The system layer supports four basic functions: the synchronization of multiple compressed streams on playback, the interleaving of multiple compressed streams into a single stream, the initialization of buffering for playback start up, and time identification.

10.1.2 Video Coding (11172-2)

The MPEG1 video coding standard specifies the coded representation of video for digital storage media and specifies the decoding process. The representation supports normal speed forward playback as well as special functions such as random access, fast play, fast reverse play, normal speed reverse playback, pause, and still procedures. This international standard is compatible with standard 525- and 625-line television formats and provides flexibility for use with personal computer and workstation displays.

This international standard is primarily applicable to digital storage media supporting a continuous transfer rate up to about 1.5 Mbps, such as compact disc, digital audio tape, and magnetic hard disks. The storage media may be directly connected to the decoder or via communications means such as busses, LANs, or telecommunications links. This international standard is intended for noninterlaced video formats having approximately 288 lines of 352 pels and picture rates around 24 Hz to 30 Hz. The objective of the basic MPEG1 coding algorithm is to provide VCR picture quality that is similar to H.261, that is, 8×8 DCT, interframe prediction, motion compensation.

The MPEG1 coding scheme is very similar to that of ITU-T H.261. The major difference between the two is that MPEG1 allows bidirectional interpolation of frames (B frames). A video sequence is divided into *groups of pictures* (GOPs) as shown in Figure 10.1. There are three possible types of pictures in a GOP, specifically, I-Picture, P-picture, and B-picture. Coding of I- and P-pictures is similar to the scheme used for the JPEG and H.261 standards, that is, intraframe and liner prediction interframe techniques with DCT. Coding of B-pictures is slightly different from that of the P-pictures. In P-pictures, motion

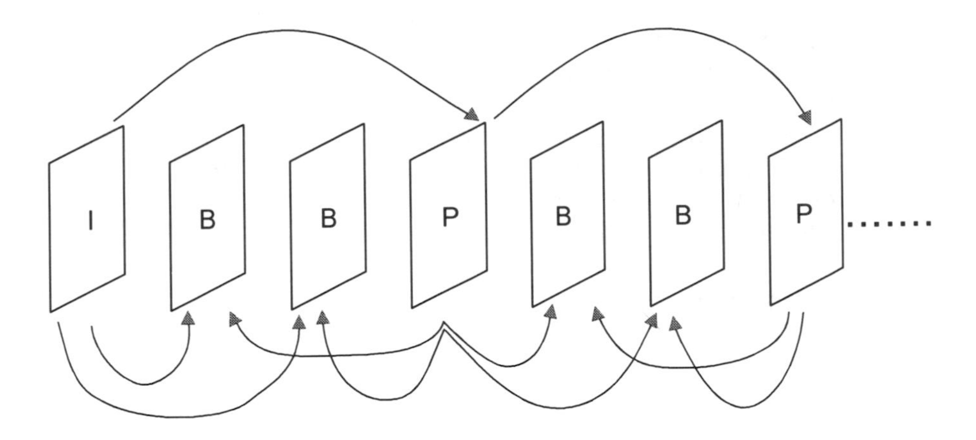

Figure 10.1 MPEG frame coding.

prediction is obtained from some previous frame only, that is, forward prediction. However, in B-pictures, information from both previous and future frames is used for prediction, that is, bidirectional interpolation.

10.1.3 Audio Coding (11172-3)

The MPEG1 audio coding standard specifies the coded representation of high-quality audio for storage media and the method for decoding of high-quality audio signals. It is compatible with the current formats (compact disc and digital audio tape) for audio storage and playback. This representation supports normal speed playback. This standard is intended for application to digital storage media providing a total continuous transfer rate of about 1.5 Mbps for both audio and video bit streams, such as CD, DAT, and magnetic hard disc. The storage media may either be connected directly to the decoder or via other means such as communication lines and the MPEG systems layer. This standard is intended for sampling rates of 32 kHz, 44.1 kHz, 48 kHz and 16-bit PCM input/output to the encoder/decoder.

10.2 MPEG2 (13818)

10.2.1 MPEG2 System (13818-1)

The systems part of MPEG2 addresses the combining of one or more elementary streams of video and audio, as well as other data, into single or multiple streams that are suitable for storage or transmission. System coding is specified in two possible forms, namely, the transport stream and the program stream. Each is optimized for a different set of applications. Both the transport stream and program stream defined in this international standard provide coding syntax that is necessary and sufficient to synchronize the decoding and presentation of the video and audio information, while ensuring that data buffers in the decoders do not overflow or underflow. Information is coded in the syntax using time stamps concerning the decoding and presentation of coded audio and visual data and time stamps concerning the delivery of the data stream itself. Both stream definitions are packet-oriented.

The basic multiplexing approach for single video and audio elementary streams is illustrated in Figure 10.2. The video and audio data is encoded as described in Parts 2 and 3 of this international standard. The resulting compressed elementary streams are packetized to produce PES packets.

The program stream is analogous, and similar to, the MPEG1 Systems Multiplex. It results from combining one or more streams of PES packets, which have a common time base, into a single stream. The program stream definition can also be used to encode multiple audio and video elementary streams into multiple program streams, all of which have a common time base.

The transport stream combines one or more programs with one or more independent time bases into a single stream. PES packets made up of elementary streams that form a program share a common timebase. The transport stream is designed for use in environments where errors are likely, such as storage or transmission in lossy or noisy media. Transport Stream packets are 188 bytes in length.

10.2.2 MPEG2 Video (13818-2)

Like MPEG1, MPEG2 consists of three interrelated standards, particularly, video coding, audio coding, and system. Again like MPEG1, MPEG2 is a generic toolkit designed for a wide range of applications such as storage/retrieval, broadcast TV (contribution and distribution), satellite TV transmission, and high-definition TV. In many ways, MPEG2 is similar to MPEG1. The primary difference is that MPEG2 provides for the coding of an interlaced TV frame while MPEG1 is restricted to the coding of one field. This difference, and others, are summarized in Table 10.1. The table also compares MPEG operation with H.261 coding.

The MPEG2 toolkit standard is organized into levels and profiles as illustrated in Figure 10.4. In general, the level axis refers to picture resolution, where the 1440 level refers to the number of pixels/line in an HDTV picture. The profile axis describes the complexity of the set of tools required for the application. For example, the Simple Profile prohibits the use of B-frames. The

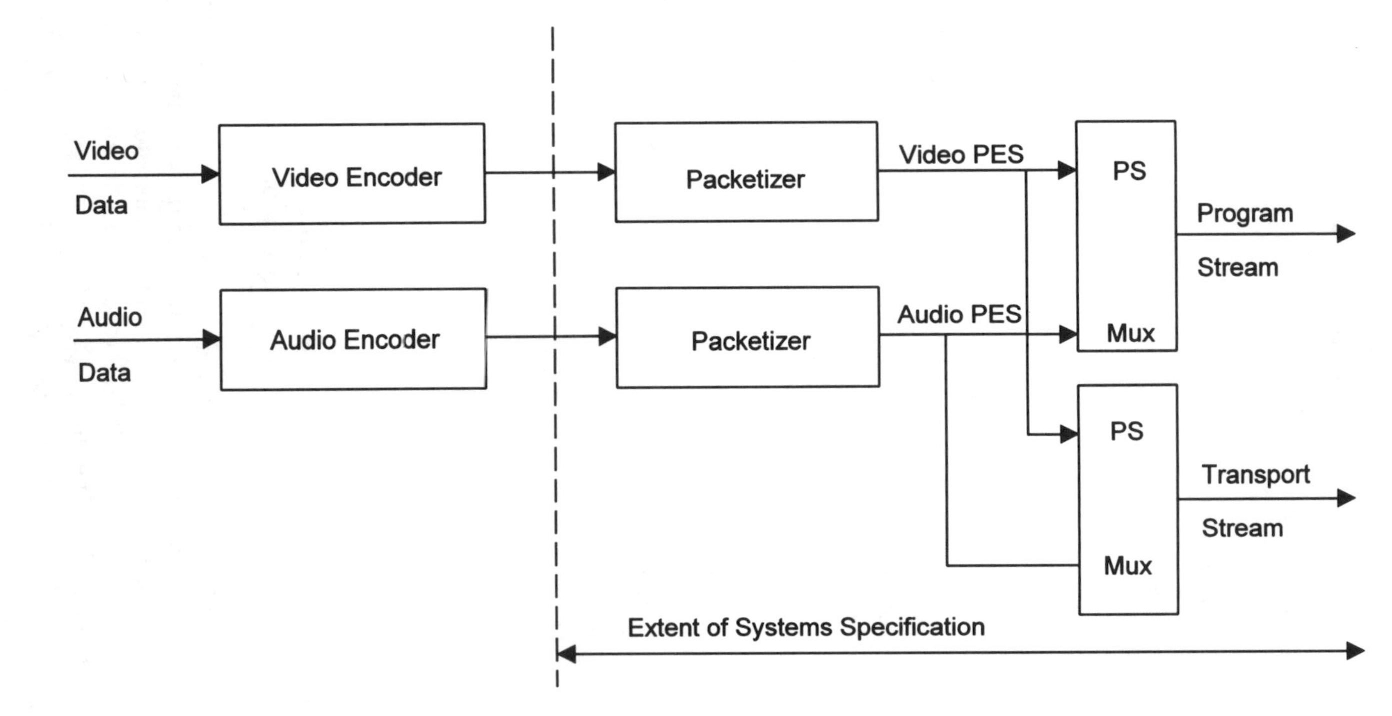

Figure 10.2 Simplified overview of ISO/IEC 13818 scope.

Table 10.1
Comparison of Video Coding Standards

	H.261	*H.263*	*MPEG1*	*MPEG2**
General standard structure	Narrow profile	Narrow profile	Generic tool kit	
Picture format	176 × 144 (Mandatory) 352 × 288 (Optional)	SQCIF, QCIF, CIF, 4CIF, 16CIF	One field	Field or frame
Quantization precision of motion vectors	One pixel	One half pixel	One half or one pixel	
B-frames/PB frame	None	PB frames	B frames An available tool	
Intraframe coding	Usually distributed	Flexible	Full frame is mandatory	
Color coding (see Figure 10.3)	4:2:0	4:2:0	4:4:4, 4:2:2, 4:2:0	
Picture structure	Group of blocks	GOB	Slice	
Dual prime (a special motion compensation mode)	No	No	No	Yes
Nominal bit rate	56 Kbps to 1,936 Kbps	Low bit rate	1.5 Mbps	4 Mbps to 20 Mbps
Applications	Interactive audiovisual services videophone videotele-conferencing	VTC/ videophone via PSTN/ mobile network	VCR	Broadcast TV contribution distribution DBS HDTV

*H.262 is a particular profile of MPEG-2.

main profile/main level (MP/ML) is the most basic and fundamental application of MPEG2 employing a resolution and complexity appropriate for conventional broadcast TV.

MPEG2 MP frame-picture coding allows three motion compensation modes (that is, frame, field, and dual prime) to be adaptively selected for each macroblock in a P- or B-picture. The frame motion compensation mode is the same as the prediction mode in MPEG1 where the odd and even lines of a macroblock are predicted as one unit using one motion vector. Field motion compensation allows the odd and even lines (fields) of a macroblock to be

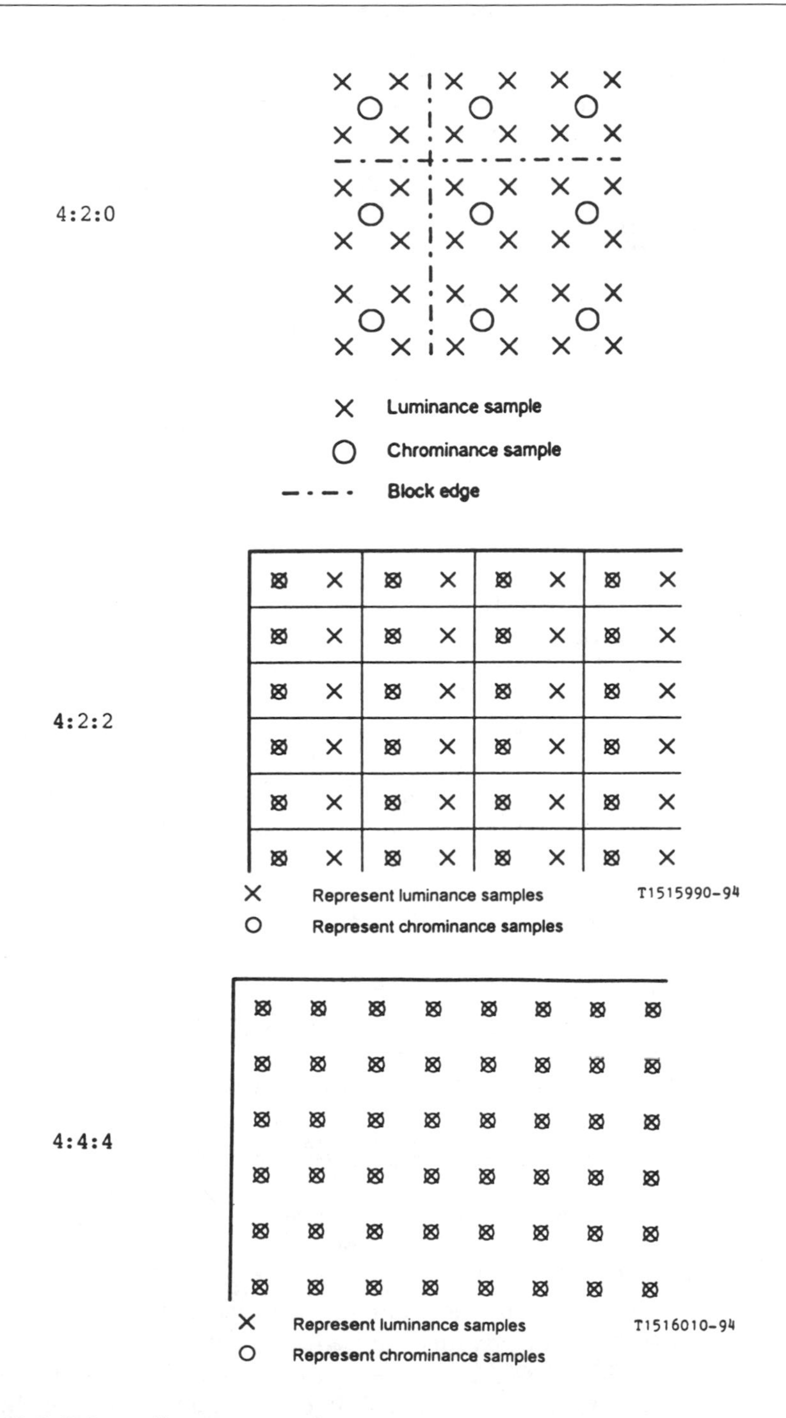

Figure 10.3 Color coding representations.

Levels						
	High Level		X			X
	High - 1440		X		X	X
	Main Level	X	X	X		X
	Low Level		X	X		
		Simple Profile	Main Profile	SNR Scalable Profile	Spatially Scalable Profile	High Profile

PROFILES (TOOL COMPLEXITY)

Figure 10.4 MPEG2 structure.

independently predicted using two separate motion vectors. Dual prime is a field-based prediction that allows interpolation of two reference fields using one motion vector and a correction vector. This mode is only allowed in a sequence with no B-pictures to reduce memory bandwidth requirement. MPEG2 also allows DCT to be performed on a frame-format block or a field-format block. The latter is created by separating the odd and even lines of a macroblock. These motion prediction and frame/field DCT options allow MPEG2 to handle interlaced video coding efficiently, which is not possible with MPEG1.

10.3 MPEG4 (14496)

In the past, standardized coding schemes such as H.320 and MPEG1, 2 each have defined a bit-precise syntax for efficiency and ease of decoding. These schemes also have been able to include a limited degree of flexibility, most notably in the form of *profiles* in MPEG2. However, they still code *video*, that is, a regular time sequence of two-dimensional video frames and audio samples. See Figure 10.5. In contrast, MPEG4 will probably standardize a generic communication language for audiovisual objects.

MPEG4 will not standardize a single algorithm. No such algorithm exists when considering the range of functionalities and applications to be addressed. Also, there is no need to standardize a single algorithm if cost-effective systems can be built to switch between algorithms or even learn new ones. The latter capability also permits future advances in coding techniques to be included in the standard. So MPEG4 may establish an extendible set of coding tools that can be combined in various ways to make algorithms, and the algorithms can

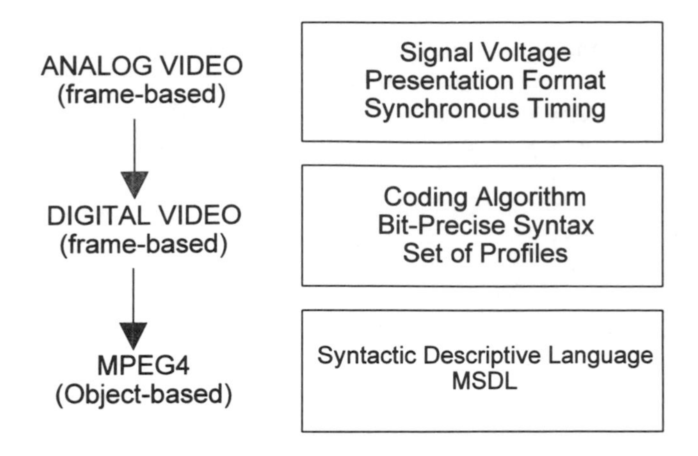

Figure 10.5 Evolution of audiovisual standards.

be customized for specific applications to make profiles. This is illustrated in Figure 10.6.

Tools, algorithms, and profiles are coding objects and consist of an independent core and a standard interface. The standard interface guarantees the coding objects can interwork, and the independent core permits proprietary techniques to be invented and made available within the standard. This is

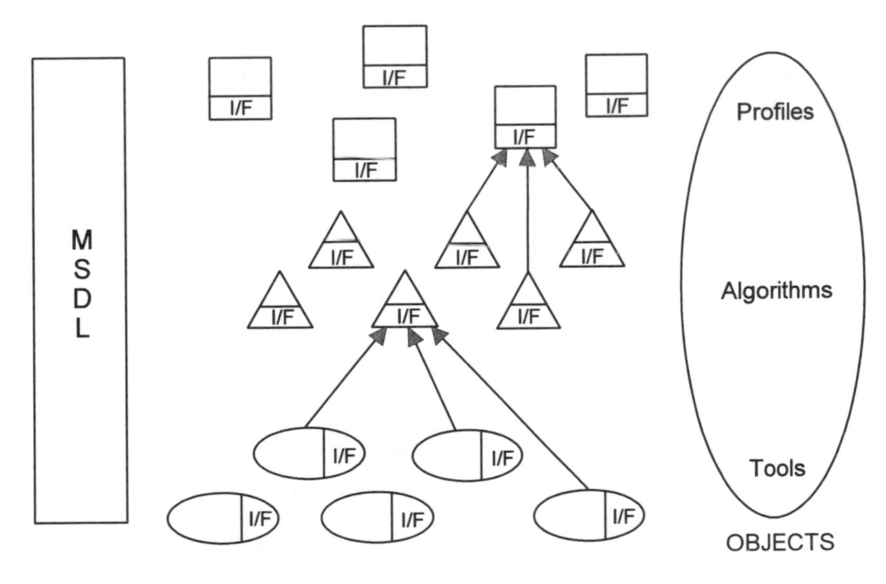

Figure 10.6 Structure of MPEG4. (*Note:* I/F is an *interface.*)

analogous to the situation in computer software applications, where independent software vendors can develop and market products that are guaranteed to run, provided they are compatible with the *application program interface* (API). In this sense, MPEG4 will be the API for the coded representation of audiovisual data.

The "glue" that binds the coding objects together is the *MPEG4 Syntactic Descriptive Language* (MSDL), which is comprised of the several key components such as the definition of the coding object interface previously noted, a mechanism to combine coding objects to construct coding algorithms and profiles, and a mechanism for downloading new coding objects. The current thinking is for coded data objects themselves to be described by the coding objects. Collectively these components define a syntax for MPEG4, and the fourth component of MSDL is a set of rules for parsing this syntax.

This structure implies a multiphase transmission of MPEG4 data. At the beginning of an exchange between a user and a data base or between two users, there is a configuration phase, during which the coder and decoder determine the coding objects to be used, their configuration, and whether or not both of them have all the required objects. If not, there may be a learning phase, during which coding objects are downloaded. Finally, there is the transmission phase for the communication of the data, which must of course be bit-efficient. The process is illustrated in Figure 10.7.

10.4 JPEG CODING ALGORITHM

The JPEG is an ISO/ITU working group that developed an international standard ("Digital Compression and Coding of Continuous-Tone Still Images") for general-purpose, continuous-tone (gray scale or color), still-image compression. The aim of the standard algorithm is to be general purpose in scope to support a wide variety of image communications services. JPEG reports jointly to both

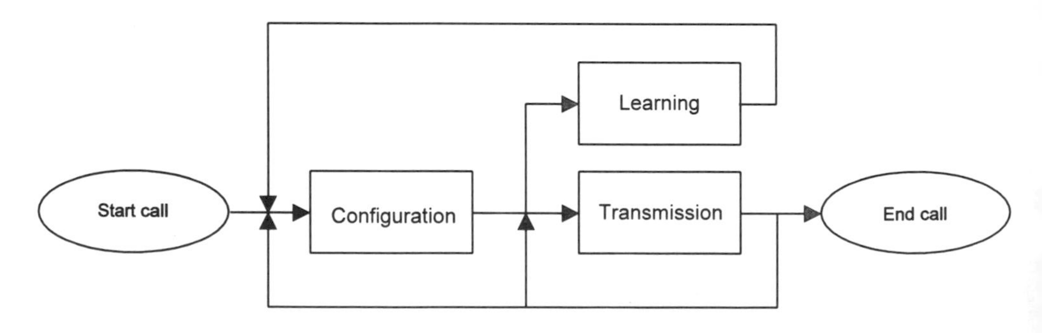

Figure 10.7 MPEG4 communication phases.

the ISO group responsible for Coded Representation of Picture and Audio Information (ISO/IEC JTC1/SC2/WG8) and to the ITU Special Rapporteur group for Common Components for Image Communication (a subgroup of ITU SGX). This dual reporting structure is intended to ensure that the ISO and the ITU reduce compatible image compression standards.

The JPEG draft standard specifies two classes of encoding and decoding processes, lossy and lossless processes. Those based on the DCT are lossy, thereby allowing substantial compression to be achieved while producing a reconstructed image with high visual fidelity to the encoder's source image. The simplest DCT-based coding process is referred to as the baseline sequential process. It provides a capability that is sufficient for many applications. There are additional DCT-based processes that extend the baseline sequential process to a broader range of applications. In any application environment using extended DCT-based decoding processes, the baseline decoding process must be present to provide a default decoding capability. The second class of coding processes is not based upon the DCT and is provided to meet the needs of applications requiring lossless compression (e.g. medical x-ray imagery.) These lossless encoding and decoding processes are used independently of any of the DCT-based processes.

10.4.1 The Baseline System

Baseline system is the name given to the simplest image coding/decoding capability proposed for the JPEG standard. It consists of techniques that are well known to the image coding community, including 8 × 8 DCT, uniform quantization, and Huffman coding. Together these provide a lossy, high-compression image coding capability, which preserves good image fidelity at high compression rates. The baseline system provides sequential build-up only.

The baseline system codes an image to full quality in one pass and is geared toward line-by-line scanners, printers, and Group 4 facsimile machines. Typically, the process starts at the top of the image and finishes at the bottom; allowing the recreated image to be built up on a line-by-line basis. One advantage is that only a small part of the image is being buffered at any given moment. Another feature stipulates that the recreated image need not be an exact copy of the original—the idea being that an almost indistinguishable copy of the original is just as good as an exact copy for most purposes. By not requiring exact copies, higher compression, which translates into lower transmission times, can be realized. Together these features are known as lossy sequential coding or transmission.

Figure 10.8 shows the main procedures for all encoding processes based on the DCT. It illustrates the special case of single component image (as opposed to multiple component color images); this is an appropriate simplification for

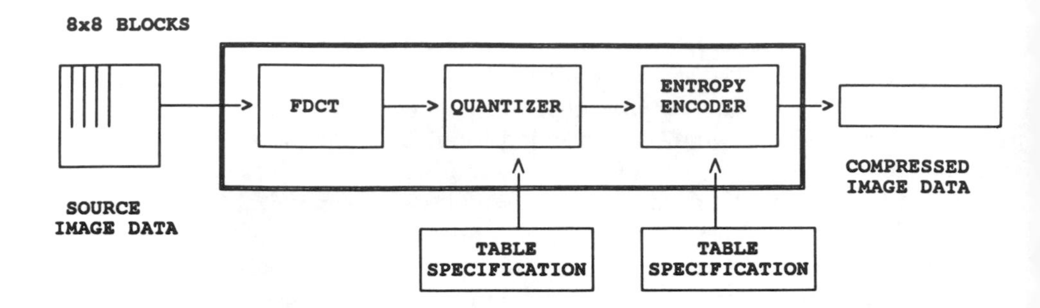

Figure 10.8 DCT-based encoder simplified diagram.

overview purposes because all processes specified in this International Standard operate on each image component independently.

In the encoding process, the input component's samples are grouped into 8×8 blocks, and each block is transformed by the forward DCT (FDCT) into a set of 64 values referred to as DCT coefficients. One of these values is referred to as the DC coefficient, and the other 63 as the AC coefficients.

Each of the 64 coefficients is then quantized using one of 64 corresponding values from a quantization table (determined by one of the table specifications shown in Figure 10.8). No default values for quantization tables are specified in this International Standard; applications may specify values that customize picture quality for their particular image characteristics, display devices, and viewing conditions.

After quantization, the DC coefficient and the 63 AC coefficients are prepared for entropy encoding. The previous quantized DC coefficient is used to predict the current quantized DC coefficient, and the difference is encoded. The 63 quantized AC coefficients undergo no such differential encoding but are converted into a one-dimensional zig-zag sequence that is common for DCT coding.

All of the quantized coefficients are then passed to an entropy encoding procedure, which compresses the data further. Since Huffman coding is used in the baseline system, Huffman table specifications must be provided to the encoder, as indicated in Figure 10.8.

Huffman coding has two forms, namely, fixed and adaptive. Fixed Huffman coding assumes that coding tables can be generated in advance from test images and then used for many images. In adaptive Huffman coding, the encoder analyzes an image's statistics before coding and devises Huffman tables tailored to that image. These tables are then transmitted to the decoder. Then the image is coded and transmitted. Upon receipt, the decoder can reconstruct the image using the previously transmitted, tailor-made, Huffman tables.

Figure 10.9 shows the main procedures for all DCT-based decoding processes. Each step shown performs essentially the inverse of its corresponding main procedure within the encoder. The entropy decoder decodes the zigzag sequence of quantized DCT coefficients. After dequantization, the DCT coefficients are transformed to an 8 × 8 block of samples by the inverse DCT (IDCT). For DCT-based processes, two alternative sample precisions are specified, namely, either 8 bits or 12 bits per sample. The baseline process uses only 8-bit precision.

10.4.2 Extended System

Extended system is the name given to a set of additional capabilities not provided by the baseline system. Each set is intended to work in conjunction with, and to build upon, the components internal to the baseline system in order to extend its modes of operation. These optional capabilities, which include arithmetic coding, progressive build-up, *progressive lossless* coding, and others, may be implemented singly or in appropriate combinations.

Arithmetic coding is an optional, "modern" alternative to Huffman coding. Because the arithmetic coding method chosen adapts to image statistics as it encodes, it generally provides 5 to 10% better compression than the Huffman method chosen by JPEG. This benefit is balanced by some increase in complexity.

Progressive build-up, the alternative to sequential build-up, is especially useful for human interaction with picture databases over low-bandwidth channels. For progressive coding, a coarse image is sent first; then refinements are sent, improving the coarse image's quality until the desired quality is achieved. This process is geared toward applications such as image databases with multiple resolution and quality requirements, freeze-frame teleconferencing, photovideotex over low-speed lines, and database browsing. There are three different,

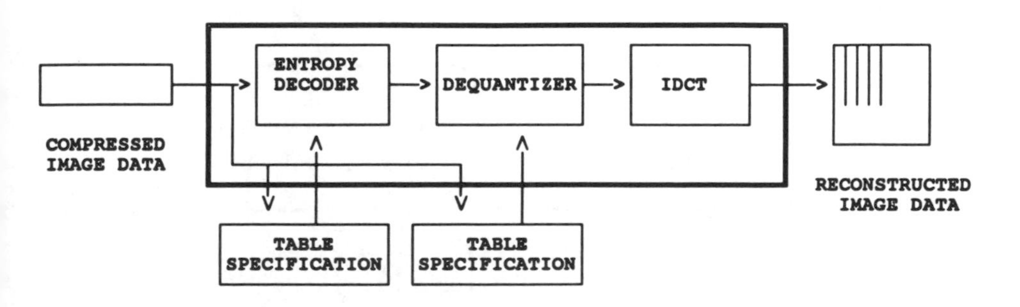

Figure 10.9 DCT-based decoder simplified diagram.

complementary, progressive, extensions—spectral selection, successive approximations, and hierarchical.

Progressive lossless refers to a lossless compression method that operates in conjunction with progressive build-up. In this mode of operation, the final stage of progressive build-up results in a received image that is bit-for-bit identical to the original.

The JPEG draft standard includes the requirement that the baseline system be contained within every JPEG-standard codec, which utilizes any of the extended system capabilities. In this way, the baseline system can serve as a default communications mode for services that allow encoders and decoders to negotiate. In such cases, image communicability between any JPEG sender and receiver that are not equipped with a common set of extended system capabilities is assured.

10.4.3 Lossless Coding

Figure 10.10 shows the main procedures for the lossless encoding processes. A predictor combines the values of up to three neighborhood samples (A, B, and C) to form a prediction of the sample indicated by X in Figure 10.11. This prediction is then subtracted from the actual value of sample X, and the difference is losslessly entropy-coded by either Huffman or arithmetic coding.

10.5 THE JBIG CODING ALGORITHM

In 1988 an experts group was formed to establish an international standard for the coding of bilevel images. The *Joint Bilevel Image Group* (JBIG) is sponsored by the ISO (IEC/JTC1/SC2/WG9) and the ITU (SG 8). In 1993 the JBIG finalized the standard entitled "Progressive Bilevel Image Compression Standard," and much of the material in this section is derived from this document.

The JBIG standard defines a method for compressing a bilevel image (that is, an image that like a black-and-white image has only two colors). Because

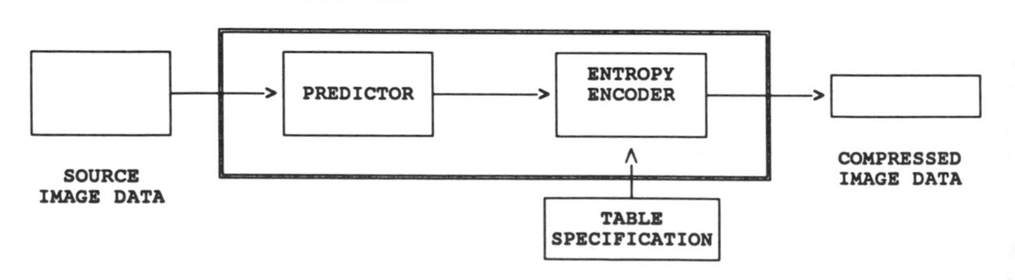

Figure 10.10 Lossless encoder simplified diagram.

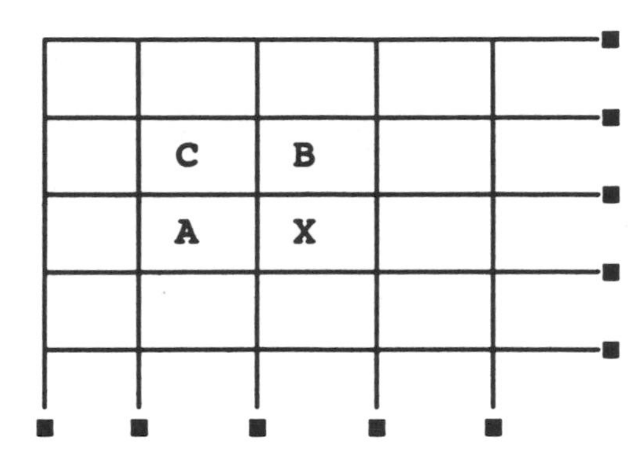

Figure 10.11 Three-sample neighborhood.

the method adapts to a wide range of image characteristics, it is a very robust coding technique. On scanned images of printed characters, observed compression ratios have been from 1.1 to 1.5 times as great as those achieved by the *Modified READ* (MMR) encoding of the ITU Recommendations T.4 (G3) and T.6 (G4). On computer generated images of printed characters, observed compression ratios have been as much as five times as great. On images with greyscale rendered by halftoning or dithering, observed compression ratios have been up to 30 times as great.

The method is bit-preserving, which means that it, like the ITU T.4 and T.6 recommendations, is distortionless and that the final decoded image is identical to the original.

The JBIG standard provides for both sequential and progressive operation. When decoding a progressively coded image, a low-resolution rendition of the original image is made available first with subsequent doublings of resolution as more data is decoded. Progressive encoding has two distinct benefits. One is that one common database can efficiently serve output devices with widely different resolution capabilities. Only that information in the compressed image file that allows reconstruction to the resolution capability of the particular output device need be sent and decoded. Also, if additional resolution enhancement is desired, for say, a paper copy of something already on a CRT screen, only the needed updating information has to be sent.

The other benefit of progressive encoding is that it provides subjectively superior image browsing (on a CRT) over low-rate and medium-rate communication links. A low-resolution rendition is rapidly transmitted and displayed, with as much resolution enhancement as desired then following. Each stage of resolution enhancement builds on the image already available. Progressive encoding makes it easy for a user to quickly recognize the image being displayed, which in turn makes it possible for that user to quickly interrupt the transmission of an unwanted image.

Although the primary aim of this recommendation is bilevel image encoding, it is possible to effectively use this standard for multilevel image encoding by simply encoding each bit plane independently as though it were itself a bilevel image. When the number of such bit planes is limited, such a scheme can be very efficient while at the same time providing for progressive buildup in both spatial refinement and greyscale refinement.

Glossary

Adaptive differential pulse code modulation	A form of differential pulse code modulation that uses adaptive quantizing.
A-law	A PCM coding and companding standard used in Europe. One common use is for digital voice communications.
Analog-to-digital (A-D) conversion	The conversion of an analog signal to a digital signal.
Asynchronous	A transmission method in which units of data are sent one character at a time preceded and followed by start/stop bits that provide timing (synchronization) at the receive terminal.
Audiovisual	Involving both hearing and sight.
Bandwidth	The difference between the limiting frequencies within which performance of a device, in respect to some characteristic, falls within specified limits. For example, NTSC signal with video and audio requires a medium with a 6-MHz bandwidth.
Bit rate	In a bit stream, the number of bits occurring per unit time, usually expressed as bits per second.
Bit rate allocation signal (BAS)	Bit position within the frame structure of ITU recommendation H.221 that is used to transmit commands, control and indication signals, and capabilities.

Broadcast operation	The transmission of information so that it may be simultaneously received by stations that usually make no acknowledgement.
Camera	In television, an electronic device using an optical system and a light-sensitive pickup tube or chip to convert visual signals into electrical impulses.
Channel	A single unidirectional or bidirectional path for transmitting or receiving, or both, of electrical or electromagnetic signals, usually in distinction from other parallel paths.
Channel service unit (CSU)	User-owned equipment installed on customer premises at the interface between customer premises and the operating phone company to terminate a circuit. CSUs provide network protection and diagnostic capabilities.
Chrominance	The difference between a reproduced color and a standard reference color of the same luminous intensity.
CIF	See FCIF.
Classified	Any information that has been determined to require protection against unauthorized disclosure to avoid harm to the U.S. national security. The classifications TOP SECRET, SECRET, and CONFIDENTIAL are used to designate such information, referred to as *classified information.*
CODEC	Acronym for COder/DECoder. An electronic device that converts analog signals, typically video, voice, and/or data into digital form and compresses them into a fraction of their original size to save frequency bandwidth on a transmission path.
Communications	A method or means of conveying information of any kind from one person or process to other person(s) or process(es) by a telecommunication medium.
Compression	The application of any of several techniques that reduce the number of bits required to represent information in data transmission or storage, therefore conserving bandwidth and/or memory.

COMSEC equipment — Equipment designed to provide security to telecommunications by converting information to a form unintelligible to an unauthorized interceptor and by reconverting such information to its original form for authorized recipients, as well as equipment designed specifically to aid in, or as an essential element of, the conversion process. Communications security equipment is cryptoequipment, cyrptoancillary equipment, cryptoproduction equipment, and authentication equipment.

Conferencing — Programs and meetings that may be for the purpose of presenting and exchanging information, comparing views, learning, planning, and decision making. Conferences can be held in one location or conducted simultaneously at multiple locations and linked together by telecommunications systems. It includes the design and engineering of conferencing systems and telecommunications services, creation of presentation media, and the development and promulgation of policy, procedures and standards for the operation of conferencing activities, facilities, systems, and networks.

Cryptography — The principles, means, and methods for rendering plain information unintelligible and for restoring encrypted information to intelligible form.

Customer premises equipment (CPE) — Terminal and associated equipment located at a subscriber's premise and connected with the termination of a carrier's communication channel(s) at the network interface at the subscriber's premises. Excluded from CPE are over-voltage protection equipment, inside wiring, coin-operated or pay telephones, and multiplexing equipment to deliver multiple channels to the customer.

Data — Representation of facts, concepts, or instructions in a formalized manner suitable for communication, interpretation, or processing by humans or by automatic means. A representation such as characters or analog quantities to which meaning is or might be assigned.

Data circuit-terminating equipment (DCE)	The interfacing equipment sometimes required to couple the data terminal equipment (DTE) into a transmission circuit or channel and from a transmission circuit or channel into the DTE.
Data communications port	A port used for the transfer of information between functional units by means of data transmission according to a protocol.
Data encryption standard (DES)	A national standard used in the United States for the encryption of information digitally transferred using a 64-bit key. The standard, which has been set by the National Bureau of Standards, provides only privacy protection and is not recognized by NSA as providing security protection.
Data rate	In digital data communications, the speed at which data (bits in this case) is transmitted, usually expressed in bits per second.
Data terminal equipment (DTE)	Equipment consisting of digital end instruments that convert the user information into data signals for transmission or reconvert the received data signals into user information.
Data transmission	Conveying data from one place for reception elsewhere by telecommunication means.
Decode	To convert data by reversing the effects of some previous encoding.
Decoder	A device that decodes. See decode.
Desktop and individual workstation	An input/output display device with local computer power allowing an individual to perform some computational work and database access from a local or remote location. It may also have videophone and/or VTC capabilities.
Differential pulse code modulation	A process in which a signal is sampled, and the difference between each sample of this signal and its estimated value is quantized and converted by encoding to a digital signal.
Discrete signals	A signal composed of sample values uniformly spaced in time. The result of sampling a continuous signal.

Duplex operation	An operating method in which transmission is permitted, simultaneously, in both directions of a telecommunication channel.
Echo	A wave that has been reflected or otherwise returned with sufficient magnitude and delay to be perceived.
Echo attenuation	In a communication circuit (4- or 2-wire) in which the two directions of transmission can be separated from each other, the attenuation of echo signals that return to the input of the circuit under consideration.
Echo cancellation	The process of reducing echo electronically in the audio system.
Echo canceller	A device that electronically reduces echo in the audio system.
Encoder	A device that encodes. See encoding.
Encoding	The process of reforming information into a format suitable for transmission.
Encrypt	To convert plain text into an unintelligible form by means of a cryptosystem.
Encryption	The process of encrypting.
Full common intermediate format (FCIF)	A video format defined in H.261 that is characterized by 352 luminance pixels on each of 288 lines, with half as many chrominance pixels in each direction.
Forward error correction (FEC)	A system of error control for data transmission wherein the receiving device has the capability to detect and correct any character or code block that contains fewer than a predetermined number of symbols in error. (FEC is accomplished by adding bits to each transmitted character or code block using a predetermined algorithm.)
Frame	The set of all the picture elements in an image.
Full duplex	An operating method in which transmission is permitted, simultaneously, in both directions of a telecommunication channel.
Graphics	The art or science of conveying information through the use of, for example, graphs, letters, lines, drawings, and pictures.

Handshaking	The process used to establish communications parameters between two stations.
High-resolution graphics	A video system that provides better resolution than a standard home television. It is not unusual for a typewritten page to be easily read on a high-resolution monitor. Graphics with sufficient detail to be able to read a full 8 1/2- by 11-in typed page; approximately 1024 × 1024 pixels of resolution.
Input/output (I/O) device (equipment)	A device that introduces data into or extracts data from a system.
Interoperability	The condition achieved among communication-electronics systems or items of communications-electronics equipment when information or services can be exchanged directly and satisfactorily between them and/or to their users. The degree of interoperability should be defined when referring to specific cases.
Integrated services digital network (ISDN).	A project underway within the CCITT for the standardization of operating parameters and interfaces for a network that will allow a variety of mixed digital transmission services to be accommodated. Access channels include a basic rate (two 64-Kbps 'B' channels + one 16-Kbps 'D' channel) and a primary rate (23 64-Kbps 'B' channels and one 64-Kbps 'D' channel).
Key	A code that governs the encryption and decryption of information. Users must have the same key in order to decrypt each other's messages.
Lip synchronization	The relative timing of audio and video signals so that there is no noticeable lag or lead between audio and video.
Long-haul communications	Communication that permits users to convey information on a national or worldwide basis.
Luminance	The monochromatic signal used to convey brightness information.
Microphones	Devices that convert acoustic energy (sound waves) into electrical energy, to be transmitted over wire or other channels of communication. An audio transducer that converts sound pressure waves (sound energy) into electrical signals.

Monitors	See visual display unit.
Motion compensation coding	A type of interframe coding used by picture processors in the compression of video images. The process relies upon an algorithm that examines a sequence of frames to develop a prediction as to the motion that will occur in subsequent frames.
Mu-law	The PCM coding and companding standard used in Japan and North America.
Multiplexer	A device for combining two or more channels into a single channel.
Multipoint	A telecommunications system that allows each of three or more sites to both transmit signals to and receive signals from all other sites.
Network	An interconnection of three or more communication entities and (usually) one or more nodes.
NTSC Standard	Acronym for National Television Standards Committee standard. North American standard for the generation, transmission, and reception of television communications wherein the 525-line picture is the standard versus the PAL and SECAM systems using the 625-line picture. (The picture information is transmitted in AM and the sound information is transmitted in FM. Compatible with CCIR Standard M.)
P × 64	Family of five ITU Recommendations. These include H.261, H.221, H.242, H.320, and H.230.
Parity	In binary-coded systems, the oddness or evenness of the number of ones in a finite binary system.
Pixel	A picture element that contains gray scale or color information. (Gray scale is an integration of density and gives resolution in terms of amplitude.)
Plain text	Digital or analog signals that are not encrypted and from which the information can be extracted relatively easily.
Point-to-point link	A data communication link connecting only two stations.
Point-to-point transmission	Transmission between two designated stations.

Point-to-point — A communication link or transmission between only two stations.

Pulse code modulation (PCM) — A process in which a signal is sampled, and each sample is quantized independently of other samples and converted by encoding to a digital signal.

Quarter Common Intermediate Format (QCIF) — A video format defined in H.261 that is characterized by 176 luminance pixels on each of 144 lines, with half as many chrominance pixels in each direction. QCIF has one-quarter as many pixels as FCIF.

Quantization — A process in which the continuous range of values of a signal is divided into nonoverlapping subranges, a discrete value being uniquely assigned to each subrange.

Raster — A predetermined pattern of scanning lines within a display space, for example, the pattern followed by an electron beam scanning the screen of a television camera or receiver.

Resolution — A measurement of the smallest detail that can be distinguished by a sensor system under specific conditions. For video equipment, often measured in terms of pixels.

Restricted channel — A digital communications channel for which each increment of p gives a useful capacity of only 56000 bits per second, instead of 64000 bits per second. This is currently common in North America and was originally due to a ones density limitation in T1 circuits.

RGB — Acronym for red-green-blue. A connection that consists of three different signals used to carry the red, green, and blue elements of a color image. Since the image is unencoded, it results in higher resolution and picture clarity than that allowed by NTSC video (which contains composite, encoded color information). Three different lines are needed for connection instead of the one for NTSC signals.

RJ-11 — Acronym for Registered Jack number 11. It is the standard modular phone jack for the United States.

RS-232	A serial interface standard for transmission of unbalanced signals between a variety of computer, media, and multimedia peripherals. It transmits at a maximum of 19.2 Kbps and uses a 25-pin connector.
RS-422	A serial interface standard for transmission of balanced and unbalanced signals between a variety of higher end computer, media, and multimedia peripherals. It allows a maximum data rate of 10 Mbps.
RS-449	A serial interface standard for transmission of balanced and unbalanced signals between a variety of higher end computer, media, and multimedia peripherals. It allows a maximum data rate of 10 Mbps and uses a 37- or 9-pin connector.
Sampling rate	The number of samples taken per unit time; the rate at which signals are sampled for subsequent use, such as modulation, coding, quantization, or any combination of these functions.
Security	The condition achieved when designated information, material, personnel, activities, and installations are protected against espionage, sabotage, subversion and terrorism as well as against loss or unauthorized disclosure. The term is also applied to those measures necessary to achieve this condition and to the organizations responsible for those measures. With respect to classified matter, it is the condition that prevents unauthorized persons from having access to official information that is safeguarded in the interests of national security.
Simplex operation	Operation method in which transmission occurs in only one preassigned direction.
Sub-band adaptive differential pulse code modulation (SB-ADPCM)	A process that splits the audio frequency band into two sub-bands (higher and lower), and the signals in each sub-band are encoded using ADPCM.
Synchronous	A process where the information and control characters are transmitted at even intervals in order to preserve continuity (synchronization) within a data communications system.

Telecommunication	Any transmission, emission, or reception of signs, signals, writings, images, and sounds or information of any nature by wire, radio, visual, or other electromagnetic systems.
Teleconferencing	Generally, the transmission of audio and/or video communications of a conference such that two or more locations are connected and can function in the live exchange of information.
Teleconferencing system	A collection of equipment and integral components (customer premise equipment and facilities) required to process teleconferencing programs and control data, less network interface devices.
Teletraining	(Also known as distance learning, teleseminar, or electronic classroom.) The use of teleconferencing point-to-point or multipoint to provide interactive remote site training.
Terminal equipment	A device or devices connected to a network or other communications system used to receive or transmit data. It usually includes some type of I/O device.
Toll quality	(3-kHz analog bandwidth) ordinary telephone voice quality.
User	A person, organization, or other entity that employs the services provided by a telecommunication system for transfer of information to others.
Video	That portion of a signal that is related to moving images.
Videoconference	See video teleconferencing.
Videoconferencing	See video teleconferencing.
Video CODEC	See CODEC.
Video teleconferencing	Two-way electronic form of communications that permits two or more people in different locations to engage in face-to-face audio and visual communication. Meetings, seminars, and conferences are conducted as if all of the participants are in the same room.
Video telephony	Relating to video phones and video teleconferencing.

Visual display unit — A device with a display screen, usually equipped with a keyboard, for example, a cathode ray tube display, light-emitting diode display, liquid crystal display, or plasma panel.

Wideband — That property of a circuit having a bandwidth greater than 4 kHz. In the case of wideband audio, G.722 specifies a bandwidth of 7 kHz.

Windowing — Capability to divide the video display into two or more separate regions with displays from different sources in each region. For example, one window could display data, another one motion video of the remote site, another one graphics, and another one motion video of the home site.

Acronyms and Abbreviations

The acronyms and abbreviations used in this document are as follows.

ANSI	American National Standards Institute
ASCII	American Standard Code for Information Interchange
BAS	bit rate allocation signal
bps	bits per second
CODEC	coder-decoder
CIF	common intermediate format
COMSEC	communications security
crypto	cryptographic
CSU	channel service unit
CTS	Clear to Send
dBm	decibel referred to 1 milliwatt
DCE	data communication equipment
DCT	discrete cosine transform
DES	Data Encryption Standard
DSU	data service unit
DTE	data terminal equipment
EIA	Electronic Industries Association
FAS	frame alignment signal
FCIF	full common intermediate format
FEC	forward error correction
Hz	Hertz
I/O	input/output
ISDN	integrated services digital network
ISO	International Organization for Standardization
JPEG	Joint Photographic Experts Group
Kbps	kilobits per second

LATA	local access and transport area
KHz	kilohertz
Mbps	megabits per second
MIL-STD	military standard
MPI	minimum picture interval
ms	milliseconds
N/A	not applicable
NATO	North Atlantic Treaty Organization
NTSC	National Television Standards Committee
PAL	phase alteration by line
QCIF	quarter common intermediate format
RGB	red-green-blue
SB-ADPCM	sub-band adaptive differential pulse-code modulation
SECAM	sequence color over memoire
VTC	video teleconferencing

About the Author

Richard A. Schaphorst, the founder and president of Delta Information Systems, Inc., obtained his B.S.E.E. degree from Lehigh University and worked at Philco Ford Aerospace in the area of image communications prior to founding Delta. In recent years, he has been one of the key leaders in the development of standards for facsimile and teleconferencing. He is a member of, and contributor to, the ANSI standards committee T1A1.5. He was recently designated by the UN/ITU as the leader (Rapporteur) to develop the H.324 series of standards for videoconferencing/videophone over the public switched telephone and mobile networks.

Mr. Schaphorst was one of three Core Members of the U.S. delegation to the Specialists Group for Coding the Visual Telephony, which developed the H.261 Recommendation for the codec operating at P × 64 Kbps ISDN rates. Mr. Schaphorst is one of two coordinators for the U.S. delegation to the ITU-T Group for ATM Video Coding.

He has published articles and given numerous presentations on the subject of image communication and standards. Recent publications include an IEEE Press offering entitled "Teleconferencing" and a book (co-authored with Dennis Bodson) entitled *Digital Fax Technology and Applications.*

Mr. Schaphorst has served as chairman of the Facsimile Equipment and System Standards Committee (TR-29) of the EIA and has been awarded three patents in the area of image processing.

Index

AAL. *See* ATM adaptation layer
Absolute category rating, 70
ACC. *See* Audio codec capabilities
Access, basic and primary, 109–10
AC coefficients, 170
ACELP. *See* Algebraic codebook excitation linear prediction
Acoustics, 15
ACR. *See* Absolute category rating
Adaptation layer, 120, 123, 132
Adaptive Huffman coding, 170
Administration, VTC system, 18
ADPCM, 68–69, 114
Affine map, 50–52
AGC protocol, 153–54
AL. *See* Adaptation layer
Algebraic codebook excitation linear prediction, 114
Alliance for Telecommunications Industry Solutions, 87
Ambient noise, 12, 15
American National Standards Institute, 87
Analysis filter, 49
ANSI. *See* American National Standards Institute
API. *See* Application Programming Interface
Application Programming Interface, 86
Arithmetic coding, 60–61, 171
Asymmetric vector quantization, 47
Asynchronous transfer mode, 129, 131–34
ATIS. *See* Alliance for Telecommunication Industry Solutions
ATM. *See* Asynchronous transfer mode
ATM adaptation layer, 132
Audio bridge, 15
Audio codec capabilities, 137–38
Audio coding, 105–6
  G.723.1 standard, 111
  H.323 standard, 143, 145
  MPEG standard, 161
Audiographics Teleconferencing Standards, Inc., 84
Audio stream, 112
Audio systems, 14–15
  quality of, 22–23
Audiovisual coding
  AV.253 standard, 22
  AV.324/M videophone, 111
  MPEG4 standard, 166–68

Bandwidth, 23
BAS. *See* Bit-rate allocation signal
Baseline system, JPEG, 169–71
Basic access, 109
Basic rate interface, 21–22
B channel, 109–10
B code, 58–59
B picture, 117, 160–61, 164, 166
Bidirectional interpolation, 160
Bilevel image, 172–74
Binary file transfer, 158
B-ISDN. *See* Broadband integrated services digital network
Bit exact standards, 67
Bitmap, 157
Bit-preserving method, 173
Bit rate
  basic access, 109
  compression coding, 29
  primary access, 109–10

Bit rate (continued)
- quality of, 21–22
- speech coder, 63
- system complexity, 7

Bit-rate allocation signal, 104–5
Bit-rate reduction, 34
Bit stream, 66, 160
Block coding, 63–64, 99–101
Blocking, 25
Blurriness, 25
BRI. *See* Basic rate interface
Brightness distortion, 34
Broadband integrated services digital network
- and N-ISDN, 138
- constant bit rate coding, 133
- guaranteed connection, 138–40
- high-resolution systems, 134–38
- nonguaranteed connection, 139–47
- overview of, 129–33
- variable bit rate coding, 133
- visual telephone adaptation, 133–34

C&I. *See* Control and indication
Call, 146
Cameras, 10, 12–13
Carrier sense multiple access with collision detection, 138
CATS. *See* Consortium for Audiographics Teleconferencing Standards, Inc.
CBR. *See* Constant bit rate
CCITT. *See* International Telephone and Telegraph Consultative Committee
CCR. *See* Comparison category rating
CD. *See* Collision detection
Chair control, 107–8
Channel coder, 29
CIC. *See* Control & indication capabilities
CIF. *See* Common intermediate format
Cluster, vector, 47
Coded Representation of Picture and Audio Information, 169
Coding
- arithmetic, 59, 171
- block, 63–64, 99–101
- channel, 29
- color, 165
- constant bit rate, 133
- discrete cosine transform, 33
- dual-rate, 115
- entropy, 47, 170–71
- fractal, 50–52
- H.261 standard, 92, 94–96
- interframe, 31, 49, 51, 97, 99, 102
- intraframe, 31, 97, 99
- ITU standards, 111
- JBIG algorithm, 172–74
- JPEG algorithm, 168–72
- knowledge-based, 55
- lossless, 169, 171–72
- lossy, 169
- object-based, 52–55
- predictive, 34–37
- pulse code modulation, 33–34
- region-based, 55
- semantic, 55
- sequential, 173
- shift, 58–59
- source, 29
- transform, 37–39
- transform coefficient, 39–45
- variable bit rate, 133
- variable length code, 31, 33
- waveform, 33
- wavelet, 47–50
- *See also* Audio coding; Audiovisual coding; Discrete cosine transform; Speech cod[illegible] Variable-length coding

Collage theorem, 50
Collision detection, 138
Color coding, 165
Comfort noise, 66
Comma code, 57–58
Command message, 118
Common Components for Image Communication, 169
Common intermediate format, 20, 94–96, 116, 134, 137, 144–45
Comparison category rating, 70
Complexity, speech coder, 64
Compression
- discrete cosine transform, 31
- JBIG coding ratios, 173
- lossless, 31, 33, 47, 172
- picture, 19, 21, 25
- *See also* Video compression

Compression layer, 160
Conditional variable-length code, 59
Conference control. *See* Generic conference control
Conference room. *See* Customized conference room
Conjugate structure algebraic codebook excitation linear prediction, 70–71

Consensus, International
Telecommunications Union, 74
Constant bit rate, 133
Continuous presence, 107
Control & indication capabilities, 137
Control and indication, 105
Control channel, 117, 147
Control stream, 113
Control units
H.323 standard, 143, 145
H.245 standard, 117–20
infrared, 16–17
wired, 16
Convergence sublayer, 134
Cosinusoidal transform, 39
Cost
electronic highway width, 18
travel, 6
VTC estimate, 26–27
VTC/VP systems, 1–2
CPE. *See* Customer premises equipment
CS. *See* Convergence sublayer
CS-ACELP. *See* Conjugate structure algebraic codebook excitation linear prediction
CSMA/CD. *See* Carrier sense multiple access with collision detection
Customer premises equipment, 86
Customized conference room, 7–9, 11

Data channel, 123–25
H.323 standard, 143, 145
Data element, 2
Data stream, 113
D channel, 109–10
DCT. *See* Discrete cosine transform
Decision making, 5–6
Delay, speech coder, 63–64
Delta modulation, 36
Desktop system, 7–9
Detail signal, 49
DFT. *See* Discrete Fourier transform
Differential pulse code modulation, 33, 36–37
Digital circuit multiplication, 69
Digital signal processing, 64, 70
Digital speech interpolation, 66
Discrete cosine coefficient, 170
Discrete cosine transform, 31, 33, 39–42, 45, 95, 100, 160, 166, 169–70
Discrete cosine transform coefficient, 171
Discrete Fourier transform, 38
Display system, 13–14

Distortion
H.261 coding, 52–53
G.711 standard, 65
quantization, 49
transform coding, 44–45
vector cluster, 47
Domain, 152
DSI. *See* Digital speech interpolation
DSP. *See* Digital signal processing
Dual-rate coder, 115

Echo cancellation/suppression, 15, 23–24
Echo gating, 24
ECSA. *See* Exchange Carriers Standards Association
Electronic highway, 18
Encoding algorithm, 21
End of block code, 42, 101
Endpoints, 142, 145, 152, 154
defined, 146
Endpoint-to-endpoint signaling, 146
End-to-end international connection, 65
End-to-end service, 144
Entropy coding, 47, 170–71
See also Variable length code
Entropy decoding, 171
EoB code. *See* End of block code
Error control, 24
Ethernet local area network, 139–40
Exchange Carriers Standards Association, 87
Extended system, JPEG, 171–71

FAS. *See* Frame alignment signal
Fast Ethernet, 140
Fatigue, travel, 6
FDCT. *See* Forward discrete cosine transform
FDDI, 140
Filters, wavelet, 47–50
Finite length, 49
Fixed Huffman coding, 170
Forward discrete cosine transform, 170
FPLMTS. *See* Future Public Land Mobile Telecommunication Service
Fractal, 31
Fractal coding, 50–52
Frame alignment signal, 103–4
Frame erasure concealment, 66
Frame rate, 21
Frame synchronous control, 105
Front-end clipping, 66
Full CIF, 29, 31
Full duplex channel, 7, 24

Full-frame resolution, 134–38
Future Land Mobile Telecommunication Service, 70
Future Public Land Mobile Telephone Service, 69

G.711, 10, 22–23, 65, 69, 105–6, 137
G.721, 65, 67, 69
G.722, 10, 22, 23, 68, 102, 105–6, 137, 145
G.723, 4, 63, 71, 111–15, 127, 145
G.726, 70, 114–15
G.727, 68
G.728, 10, 22–23, 64–65, 68–70, 105–6, 137, 145
G.729, 63, 67, 70–71, 145
Gain, compression, 47
Gatekeeper, 142, 145–46
Gateway, 146–47
GATT. *See* General Agreement on Tariffs and Trade
GCC. *See* Generic conference control
General Agreement on Tariffs and Trade, 83
Generic conference control, 150, 154–55
GK. *See* Gatekeeper
GOB. *See* Groups of blocks
GOP. *See* Groups of pictures
Graphic communications. *See* Multipoint graphic communications
Groups of blocks, 96–97, 99, 125
Groups of pictures, 160
GSTN, 111, 125, 140
Guaranteed quality of service, 138–39
GW. *See* Gateway

H.120, 91
H.130, 91
H.221, 8, 69, 92, 103–4
H.223, 4, 111–13, 116, 120–23, 127
H.224, 107, 124
H.225, 143–44
H.230, 8, 92, 105
H.231, 10, 107
H.233, 108
H.242, 8, 92, 103–5
H.243, 10, 107
H.245, 4, 111–14, 116–20
H.261, 8, 19–20, 24, 29, 33, 42, 52, 53, 92, 94–96, 102–3, 112, 116–17, 137
  compared to H.263, 117
H.262, 137
H.263, 4, 33, 111–13, 115–17
H.280, 106
H.310, 4, 112, 134–38
H.320, 2, 8, 67, 69, 86–87, 91–94, 107, 115
H.321, 4, 133–35
H.322, 4, 138–39, 141
H.323, 139–47
H.324, 2, 4, 68, 111–14, 125, 127
Half-duplex transmission, 24
Handshake protocol, 104
H channel, 109–10
High-pass filter, 48
High-rate coder, 114–15
High-resolution system, 134–38
High-speed data, 106
Hollywood squares, 107
HSD. *See* High-speed data
Huffman code, 60, 169–70

IDCT. *See* Inverse discrete cosine transform
IEC. *See* International Electrotechnical Commission
Image retention, 25
IMTC. *See* International Multimedia Teleconferencing Consortium
Indication message, 118
Information stream, 112–13
Infrared control unit, 16–17
Initialization, 160
Integrated services digital network, 1, 157
Interframe coding, 31, 49, 51, 97, 99, 102
Interleaving, 160
Intermediate reference system, 69
International Electrotechnical Commission, 80
International Multimedia Teleconferencing Consortium, 84–87
International Standards Organization, 80–83
Information Network, 83
International Telecommunications Union, 2
  mission/operation of, 73–76
  speech coder recommendations, 67–71
  Telecommunication Standardization Advisory Group, 78–79
  Telecommunication Standardization Bureau, 79–80
  Telecommunication Standardization Sector, 76–78
  VTC/VP standards, 2–4, 8, 10, 19
  World Telecommunication Standardization Conference, 78
International Telephone and Telegraph Consultative Committee, 22–23
Interning, 7
Intraframe coding, 31, 97, 99

Inverse discrete cosine transform, 171
I picture, 160
IRS. *See* Intermediate reference system
IS-54, 65, 67, 115–16
ISDN. *See* Integrated services digital network
ISO. *See* International Standards Organization
ISO-Ethernet local area network, 139
ISONET. *See* International Standards Organization, Information Network
Iterated system functions, 50
ITU. *See* International Telecommunications Union
ITU-R. *See* Radiocommunications Sector
ITU-T. *See* Telecommunication Standardization Sector

JBIG. *See* Joint Bilevel Image Group; Joint Binary Imagery Group
Jerkiness, 25
Joint Bilevel Image Group, 172–74
Joint Binary Imagery Group, 159
Joint Photographic Experts Group, 33, 60, 117, 168–72
JPEG. *See* Joint Photographic Experts Group

Karhunen-Loeve transform, 38–39
KLT. *See* Karhunen-Loeve transform
Knowledge-based coding, 55

LCN. *See* Logical channel number
LD-CELP. *See* Low-delay code excited linear prediction
Level axis, 162
Linear prediction coder, 70
Lip sync, 23
Local area network
  defined, 147
  guaranteed service, 138–39
  nonguaranteed service, 139–47
Logical channel, 112, 145, 147
Logical channel number, 120, 122
Lossless coding, 169, 171–72
Lossless compression, 31, 33, 47, 172
Lossy coding, 169
Low-bit rate multimedia communication, 111–14
  H.223 standard, 120–23
  H.263 standard, 115–17
Low-delay code excited linear prediction, 10, 65, 68–70
Low-pass filter, 48–49
Low-rate coder, 114–15
Low-speed data, 106
LSD. *See* Low-speed data

MAC. *See* Media access control
Macroblock, 97, 99, 164
Main level, 137, 164
Main profile, 137, 164
Manual camera switching, 13
MC. *See* Multiplexes code
MCCOI. *See* Multimedia Communications Community of Interest
MCS. *See* Multipoint commuications service
MCU. *See* Multipoint control unit
Media access control, 138
Meetings
  number of, 6
  productive, 6–7
Messages, H.245, 118
ML. *See* Main level
MLP. *See* Multilayer protocol
MMR. *See* Modified read
Mobile radio, 125–27
Modified Read, 172
Monitors, 13–14
Morale, employee, 6
Motion compensation, 21, 95, 101–3, 164
Motion Picture Experts Group, 33, 133, 137–38, 145, 159
  audio coding, 161
  MPEG1 project, 159–61
  MPEG2 project, 159, 161–66
  MPEG4 project, 166–68
  system layer, 160
  video coding, 160–61
MP. *See* Main profile
MPEG. *See* Motion Picture Experts Group
MPEG4 syntactic descriptive language, 168
MP-MLQ. *See* Multipulse-maximum likelihood quantizer
MSDL. *See* MPEG4 syntactic descriptive language
Multilayer protocol, 106
Multimedia, defined, 150
Multimedia communications
  H.223 standard, 120–23
  H.245 standard, 117–20
Multimedia Communications Community of Interest, 84
Multiplexes code, 122
Multiplexing, 94–95, 162–63
  H.223 standard, 120–23
Multiplex layer, 120, 122–23
Multipoint binary file transfer, 158

Multipoint communications service, 152–54
Multipoint control unit, 7, 107–8, 112, 149
Multipoint graphic commuications, 149–50
  binary file transfer, 158
  generic conference control, 154–55
  infrastructure, 150–52
  multipoint communications service, 152–54
  node controller, 150
  still images, 157–58
  transport protocol stack profiles, 155–57
Multipoint teleconferencing, 15
Multipulse-maximum likelihood quantizer, 114
MUX. *See* Multiplex layer

NAC. *See* Network adaptation capabilities
Narrowband integrated services digital network, 92, 108–9
  and B-ISDN, 138
National Television Standards Committee, 19, 94
Network adaptation capabilities, 137
Network node interface, 129–30
N-ISDN. *See* Narrowband integrated services digital network
NNI. *See* Network node interface
Node controller, 150
Noise
  ambient, 12, 15
  comfort, 66
  speech coding, 66
  quantizing, 25
NTSC. *See* National Television Standards Committee
Nyquist rate, 29, 33

OBASC. *See* Object-based analysis-synthesis coding
OBC. *See* Object-based coding
Object-based analysis-synthesis coding, 53–55
Object-based coding, 52–53
  generic unknown objects, 53–55
  knowledged-based coding, 55
ODC. *See* Other data capabilities
Oscillation, detail signal, 49
Other data capabilities, 137

Packet-switched data network, 129, 155, 157
PAL. *See* Phase alteration by line
PB frame, 117–18
PCM. *See* Pulse code modulation
PDU. *See* Protocol data unit
Perfect reconstruction, 49
Performance and Signal Processing Technical Subcommittee, 89
Performance impairment, 25
Personnel, key, 6
PES packet, 162
Phase alteration by line, 19, 94
Picture format, 19
  coding standards, 96–99
  video terminal, 116
Pixel, 29
Point-to-point connections, 7
POTS, 114
P picture, 117, 160, 164
Predictive coding, 33–37, 164
Primary access, 109–10
Privacy, 108
Productivity, 6
Profile axis, 162
Program stream, 162
Progressive build-up, 171
Progressive lossless coding, 171–73
Project proposal, 27
Protocol data unit, 150, 155, 157
PSDN. *See* Packet-switched data network
PSTN. *See* Public-switched telephone network
Psychovisual perception, 47
Public-switched telephone network, 63, 111, 155
Pulse code modulation, 33–35

Q.2931 information element, 134
QCIF. *See* Quarter common intermediate format
QDU. *See* Quantization distortion unit
QoS. *See* Quality of service
Quality, speech coder, 65–66
Quality, video teleconferencing
  audio, 22–23
  bandwidth, 23
  bit rates, 21–22
  common intermediate format, 20
  echo cancellation, 23–24
  encoding algorithm, 21
  error control, 24
  frame rate, 21
  general, 18–19
  lip sync, 23
  performance impairments, 25
  picture format, 19
  quarter common intermediate format, 20
  system considerations, 19

Quality of service
guaranteed, 138–39
nonguaranteed, 139–47
Quantization, 31, 36–37, 95, 170
compression gain, 49
errors in, 44
noise in, 25
scalar, 36, 46
uniform, 169
vector, 31, 45–47
Quantization distortion unit, 65, 69
Quarter common intermediate format, 20, 22, 94–96, 116, 137, 144

Radiocommunications Sector, 76
RAM. *See* Random access memory
Random access memory, 64
RE. *See* Receive end
Read only memory, 64
Receive end, 137
Reference signal, 49
Region-based coding, 55
Region-based wavelet transform, 50
Regularity, 49
Reliability, 7
Request message, 118
Response message, 118
Ringing, detail signal, 48
Rollabout module, 7–9, 11–14
ROM. *See* Read only memory
Router, 147
RPE-LTP, 65, 67

Safety, employee, 6
SAR. *See* Segmentation-and-reassembly
SB-ADPCM. *See* Subband adaptive differential pulse code modulation
SBC. *See* Subband coder
SC. *See* Service channel
Scalar quantization, 36, 46
Scanning order, 101
Schedule, project, 27
SCN. *See* Switched circuit network
SDH. *See* Synchronous digital hierarchy
SECAM. *See* Sequence color over memoire
Security, 6, 155
Segmentation-and-reassembly, 134
Semantic coding, 55
Sequence color over memoire, 19, 94
Sequential coding, 173
Service channel, 103
Servo loop, 37
SG. *See* Study group
Shift code, 57, 58–59
SI exchange. *See* Still image exchange
Signal analyzer, 31
Signals, detail and reference, 49
Sinusoidal transform, 39
Source coder, 29, 94–95, 98
Speakerphone, 24
Specialists Group, 91
Speech coding
bit rate, 63
complexity, 64
delay, 63–64
G.722 standard, 68
G.723 standard, 113–15
G.728 standard, 68–70
G.729 standard, 70–71
quality, 65–66
validation, 66–67
Speech Quality Experts Group, 65, 70
SQCIF, 116, 144
SQEG. *See* Speech Quality Experts Group
Standards organizations, international
International Multimedia Teleconferencing Consortium, 84–87
International Standards Organization, 80–83
International Telecommunications Union, 73–80
Telecommunication Standardization Advisory Group, 78–79
Telecommunication Standardization Bureau, 79–80
Telecommunication Standardization Sector, 76–78
World Telecommunication Standardization Conference, 78
Standards organizations, U.S., 87–90
Still image compression, 168
Still image exchange, 157–158
Study group, 77
Subband adaptive differential pulse code modulation, 106
Subband coder, 68
Switched circuit network, 142
Switched display, 107
Switched telecom, 155
Synchronization, 160
Synchronous digital hierarchy, 129
Synthesis filter, 49
System considerations, 19
System layer, 160

T.120, 86–87, 124, 149–58
T.122, 152–54, 158
T.123, 155–57
T.124, 154–55, 158
T.125, 152–54
T.126, 157–58
T.127, 157–58
T.4, 124, 173
T.6, 173
T.84, 124
T1 Advisory Group, 87
T1AG. *See* T1 Advisory Group
T1-Telecommunications committee, 87–90
TE. *See* Transmit end
Team building, 7
Technical approach, 26
Telecommunication Standardization Advisory Group, 77–79
Telecommunication Standardization Bureau, 78–80
Telecommunication Standardization Sector, 76–78
Teleconferencing. *See* Video teleconferencing
Terminal
  defined, 147
  H.323 standard, 142–46
  types of, 138
Time identification, 160
Token ring local area network, 140
Token, 154
Toll quality coder, 69
Training vector set, 47
Transform coding, 37–45
Transform coefficients, 39–44
Transmission buffer, 94
Transmission coder, 94
Transmit end, 137
Transport protocol stack profiles, 155–57
Transport stream, 162
Travel costs, 6
Truncation error, 44–45
TSAG. *See* Telecommunication Standardization Advisory Group
TSB. *See* Telecommunication Standardization Bureau
Two-dimensional code, 61

Uniform quantization, 169
Unknown objects, 53–55
User selection, 107
User survey, 26

Validation, speech coding, 66–67
Variable bit rate, 133
Variable-length coding, 31, 33, 55–57, 101
  arithmetic coding, 60–61
  B code, 59–60
  comma code, 57–58
  conditional, 60
  Huffman code, 60, 169–70
  shift code, 58–59
  two-dimensional, 61
VBR. *See* Variable bit rate
VCC. *See* Video codec capabilities
Vector quantization, 31, 45–47
Video coding
  capabilities of, 137
  H.323 standard, 143–44
  MPEG standard, 160–62, 164–66
Video compression
  arithmetic coding, 60–61
  B coding, 59–60
  coding techniques, 33
  comma coding, 57–58
  conditional variable length coding, 60
  fractal coding, 50–52
  generic object coding, 53–55
  generic view of, 30–45
  Huffman coding, 60
  knowledge-based coding, 55
  object-based coding, 52–55
  overview of, 29–30
  predictive coding, 34–37
  pulse code modulation, 33–34
  shift coding, 57
  transformation techniques, 38–39
  transform coding, 37–38
  transform coefficient coding, 39–45
  two-dimensional VLC, 61
  variable-length coding, 55–61
  vector quantization, 45–47
  wavelet coding, 47–50
Videophone, 1–4, 69
Video stream, 112
Video teleconferencing
  administration, 18
  audio system, 14–15, 22–23
  bandwidth, 23
  benefits of, 5–7
  bit rates, 21–22
  cameras, 12–13
  common intermediate format, 20
  controller, 16–17

display system, 13–14
echo cancellation, 23–24
encoding algorithm, 21
error control, 24
frame rate, 21
introduction to, 1–5
lip sync, 23
overview of, 7–10
performance impairments, 25
picture format, 19
planning for, 25–27
quality/performance, 18–25
quarter intermediate format, 20
room considerations, 10–12
system considerations, 19
Visual telephone terminal, 133
VLC. *See* Variable-length coding
Voice activation, 13, 107
Voice activity detector, 66, 68
VP. *See* Videophone
VQ. *See* Vector quantization
VSELP, 65, 114
VTC. *See* Video teleconferencing

Walsh-Hadamard transform, 39
WAN. *See* Wide area network
Waveform coding, 33
Wavelet coding, 47–50
Whiteboarding, 157–58
Wide area network, 139
Wired control unit, 16
Working Party, 77
World Telecommunication Standardization Conference, 77–78
World Trade Organization, 83
WP. *See* Working Party
WTO. *See* World Trade Organization
WTSC. *See* World Telecommunication Standardization Conference

Zone, 147

The Artech House Telecommunications Library

Vinton G. Cerf, Series Editor

Advanced Technology for Road Transport: IVHS and ATT, Ian Catling, editor

Advances in Computer Communications and Networking, Wesley W. Chu, editor

Advances in Computer Systems Security, Rein Turn, editor

Advances in Telecommunications Networks, William S. Lee and Derrick C. Brown

Analysis and Synthesis of Logic Systems, Daniel Mange

An Introduction to International Telecommunications Law, Charles H. Kennedy and M. Veronica Pastor

An Introduction to U.S. Telecommunications Law, Charles H. Kennedy

Asynchronous Transfer Mode Networks: Performance Issues, Raif O. Onvural

ATM Switching Systems, Thomas M. Chen and Stephen S. Liu

A Bibliography of Telecommunications and Socio-Economic Development, Heather E. Hudson

Broadband: Business Services, Technologies, and Strategic Impact, David Wright

Broadband Network Analysis and Design, Daniel Minoli

Broadband Telecommunications Technology, Byeong Lee, Minho Kang, and Jonghee Lee

Cellular Radio: Analog and Digital Systems, Asha Mehrotra

Cellular Radio Systems, D. M. Balston and R. C. V. Macario, editors

Client/Server Computing: Architecture, Applications, and Distributed Systems Management, Bruce Elbert and Bobby Martyna

Codes for Error Control and Synchronization, Djimitri Wiggert

Communications Directory, Manus Egan, editor

The Complete Guide to Buying a Telephone System, Paul Daubitz

Computer Networks: Architecture, Protocols, and Software, John Y. Hsu

Computer Telephone Integration, Rob Walters

The Corporate Cabling Guide, Mark W. McElroy

Corporate Networks: The Strategic Use of Telecommunications, Thomas Valovic

Current Advances in LANs, MANs, and ISDN, B. G. Kim, editor

Digital Cellular Radio, George Calhoun

Digital Hardware Testing: Transistor-Level Fault Modeling and Testing, Rochit Rajsuman, editor

Digital Signal Processing, Murat Kunt

Digital Switching Control Architectures, Giuseppe Fantauzzi

Distributed Multimedia Through Broadband Communications Services, Daniel Minoli and Robert Keinath

Disaster Recovery Planning for Telecommunications, Leo A. Wrobel

Distance Learning Technology and Applications, Daniel Minoli

Document Imaging Systems: Technology and Applications, Nathan J. Muller

EDI Security, Control, and Audit, Albert J. Marcella and Sally Chen

Electronic Mail, Jacob Palme

Enterprise Networking: Fractional T1 to SONET, Frame Relay to BISDN, Daniel Minoli

Expert Systems Applications in Integrated Network Management, E. C. Ericson, L. T. Ericson, and D. Minoli, editors

FAX: Digital Facsimile Technology and Applications, Second Edition, Dennis Bodson, Kenneth McConnell, and Richard Schaphorst

FDDI and FDDI-II: Architecture, Protocols, and Performance, Bernhard Albert and Anura P. Jayasumana

Fiber Network Service Survivability, Tsong-Ho Wu

Fiber Optics and CATV Business Strategy, Robert K. Yates *et al.*

A Guide to Fractional T1, J. E. Trulove

A Guide to the TCP/IP Protocol Suite, Floyd Wilder

Implementing EDI, Mike Hendry

Implementing X.400 and X.500: The PP and QUIPU Systems, Steve Kille

Inbound Call Centers: Design, Implementation, and Management, Robert A. Gable

Information Superhighways: The Economics of Advanced Public Communication Networks, Bruce Egan

Integrated Broadband Networks, Amit Bhargava

Intelcom '94: The Outlook for Mediterranean Communications, Stephen McClelland, editor

International Telecommunications Management, Bruce R. Elbert

International Telecommunication Standards Organizations, Andrew Macpherson

Internetworking LANs: Operation, Design, and Management, Robert Davidson and Nathan Muller

Introduction to Document Image Processing Techniques, Ronald G. Matteson

Introduction to Error-Correcting Codes, Michael Purser

Introduction to Satellite Communication, Bruce R. Elbert

Introduction to T1/T3 Networking, Regis J. (Bud) Bates

Introduction to Telecommunication Electronics, Second Edition, A. Michael Noll

Introduction to Telephones and Telephone Systems, Second Edition, A. Michael Noll

Introduction to X.400, Cemil Betanov

Land-Mobile Radio System Engineering, Garry C. Hess

LAN/WAN Optimization Techniques, Harrell Van Norman

LANs to WANs: Network Management in the 1990s, Nathan J. Muller and Robert P. Davidson

Long Distance Services: A Buyer's Guide, Daniel D. Briere

Measurement of Optical Fibers and Devices, G. Cancellieri and U. Ravaioli

Meteor Burst Communication, Jacob Z. Schanker

Minimum Risk Strategy for Acquiring Communications Equipment and Services, Nathan J. Muller

Mobile Communications in the U.S. and Europe: Regulation, Technology, and Markets, Michael Paetsch

Mobile Information Systems, John Walker

Narrowband Land-Mobile Radio Networks, Jean-Paul Linnartz

Networking Strategies for Information Technology, Bruce Elbert

Numerical Analysis of Linear Networks and Systems, Hermann Kremer *et al.*

Optimization of Digital Transmission Systems, K. Trondle and Gunter Soder

Packet Switching Evolution from Narrowband to Broadband ISDN, M. Smouts

Packet Video: Modeling and Signal Processing, Naohisa Ohta

Personal Communication Systems and Technologies, John Gardiner and Barry West, editors

The PP and QUIPU Implementation of X.400 and X.500, Stephen Kille

Practical Computer Network Security, Mike Hendry

Principles of Secure Communication Systems, Second Edition, Don J. Torrieri

Principles of Signaling for Cell Relay and Frame Relay, Daniel Minoli and George Dobrowski

Principles of Signals and Systems: Deterministic Signals, B. Picinbono

Private Telecommunication Networks, Bruce Elbert

Radio-Relay Systems, Anton A. Huurdeman

Radiodetermination Satellite Services and Standards, Martin Rothblatt

Residential Fiber Optic Networks: An Engineering and Economic Analysis, David Reed

Secure Data Networking, Michael Purser

Service Management in Computing and Telecommunications, Richard Hallows

Setting Global Telecommunication Standards: The Stakes, The Players, and The Process, Gerd Wallenstein

Smart Cards, José Manuel Otón and José Luis Zoreda

Super-High-Definition Images: Beyond HDTV, Naohisa Ohta, Sadayasu Ono, and Tomonori Aoyama

Television Technology: Fundamentals and Future Prospects, A. Michael Noll

Telecommunications Technology Handbook, Daniel Minoli

Telecommuting, Osman Eldib and Daniel Minoli

Telemetry Systems Design, Frank Carden

Telephone Company and Cable Television Competition, Stuart N. Brotman

Teletraffic Technologies in ATM Networks, Hiroshi Saito

Terrestrial Digital Microwave Communications, Ferdo Ivanek, editor

Toll-Free Services: A Complete Guide to Design, Implementation, and Management, Robert A. Gable

Transmission Networking: SONET and the SDH, Mike Sexton and Andy Reid

Transmission Performance of Evolving Telecommunications Networks, John Gruber and Godfrey Williams

Troposcatter Radio Links, G. Roda

Understanding Emerging Network Services, Pricing, and Regulation, Leo A. Wrobel and Eddie M. Pope

UNIX Internetworking, Uday O. Pabrai

Videoconferencing and Videotelephony: Technology and Standards, Richard Schaphorst

Virtual Networks: A Buyer's Guide, Daniel D. Briere

Voice Processing, Second Edition, Walt Tetschner

Voice Teletraffic System Engineering, James R. Boucher

Wireless Access and the Local Telephone Network, George Calhoun

Wireless Data Networking, Nathan J. Muller

Wireless LAN Systems, A. Santamaría and F. J. López-Hernández

Wireless: The Revolution in Personal Telecommunications, Ira Brodsky

Writing Disaster Recovery Plans for Telecommunications Networks and LANs, Leo A. Wrobel

X Window System User's Guide, Uday O. Pabrai

For further information on these and other Artech House titles, contact:

Artech House
685 Canton Street
Norwood, MA 02062
617-769-9750
Fax: 617-769-6334
Telex: 951-659
e-mail: artech@artech-house.com

Artech House
Portland House, Stag Place
London SW1E 5XA England
+44 (0) 171-973-8077
Fax: +44 (0) 171-630-0166
Telex: 951-659
e-mail: artech-uk@artech-house.com